Praise for Colin Ellard

"One of the finest science writers I've ever read."
—*Los Angeles Times*

"Delightfully lucid. . . . Ellard has a knack for distilling obscure scientific theories into practical wisdom."
—*New York Times Book Review*

"[Ellard] mak[es] even the most mundane entomological experiment or exegesis of psychological geekspeak feel fresh and fascinating."
—**NPR**

"[Ellard] entertain[s] us with an explanation of the cold, hard science of navigation [and] artfully constructed exploration[s] of how our relationship to spaces plays a huge part in making us human."
—*Quill & Quire*

"Ellard writes with admirable clarity." —*Kirkus Reviews*

"Ellard makes it clear that the space around us has made us the species we are. . . . Witty and engaging and crammed with profound insights."
—**Michael Brooks**, author of *13 Things that Don't Make Sense: The Most Baffling Scientific Mysteries of Our Time*

"[Ellard] took me on a journey to places I'd never even considered before."
—**Sarah Susanka**, author of *The Not So Big House*

PLACES OF THE HEART

The Psychogeography of Everyday Life

Colin Ellard

Bellevue Literary Press
NEW YORK

First published in the United States in 2015 by
Bellevue Literary Press, New York

For information, contact:
Bellevue Literary Press
90 Broad Street
Suite 2100
New York, NY 10004
www.blpress.org

Library of Congress Cataloging-in-Publication Data
Ellard, Colin, 1958–
Places of the heart : the psychogeography of everyday life / Colin Ellard.
pages cm
ISBN 978-1-942658-00-9 (paperback)—ISBN 978-1-942658-01-6 (ebook)
1. Space—Psychological aspects. 2. Human geography. I. Title.
HM654.E45 2015
304.2—dc23
2015023418

Bellevue Literary Press would like to thank all its generous
donors—individuals and foundations—for their support.

 This publication is made possible by the New York
State Council on the Arts with the support of Governor
Andrew M. Cuomo and the New York State Legislature.

 This project is supported in part
by an award from the National
Endowment for the Arts.

Book design and composition by Mulberry Tree Press, Inc.

Bellevue Literary Press is committed to ecological stewardship in our book production practices,
working to reduce our impact on the natural environment.

♾ This book is printed on acid-free paper.

Manufactured in the United States of America.
First Edition

7 9 8 6

paperback ISBN: 978-1-942658-00-9

ebook ISBN: 978-1-942658-01-6

For Kristine

Contents

PLACES OF THE HEART

The Psychogeography of Everyday Life

Introduction

WHEN I WAS SIX YEARS OLD and before I had even thought once about how I wanted to spend my life, my father took me to visit Stonehenge. Our visit took place about fifty years ago, long before there was any kind of regulation or control of the site—not even a fence. Early one spring morning, we stood together on the empty Salisbury Plain, walked among the giant pillars of stone, and ran our hands over their smooth surfaces, saying very little to each other. No words were required. It was enough to be there. I was too young to have any clear understanding of the gulf of time that separated us from the site's creators, and my mind was uncluttered by the years of schooling and the accretion of complicated mental associations that when I was an adult would make it so much more difficult for me to be in the simple presence of a monument, allowing the feelings that it engendered to wash over me. I knew I was in the presence of something old and important and it was clear to me that whoever had shaped and raised these giant stones had meant business—the effort required to build this thing spoke volumes. I knew very little then about the mysteries surrounding Stonehenge; though my curiosity about this gradually mounted, on my first exposure to the site, questions about its purpose mattered little to me. I was preoccupied with the feelings I experienced—a sensation of smallness even greater than that normally felt by a little boy holding his father's hand in a strange place, some breath-quickening anxiety, perhaps produced by the awareness that I was stepping among stones that must have been laid out by others for some great purpose that they never intended to share with me. I wanted to move around the pillars, peer up at their peaks, explore their surfaces, while at the same time I had a deliciously creepy feeling that perhaps we shouldn't be there at all—that the giants who made this place might soon return.

My father, a worker in the building trades, probably had a different kind of experience that day. When I was very young, I was only vaguely aware of what he did for a living, but I learned enough about his work in my teen years to understand that it was hard for him to look at any kind of built structure without calculating, making a mental inventory of the sizes and shapes of the materials that had been used, and thinking about the sturdiness of a structure and its ability to stand up to the elements and to simple human use. My father was a quantity surveyor, and it was his job to understand measurement, cost, and value in an architect's blueprint and to participate in the work that would ensure that the final building both lived up to the architect's vision and came in at or under budget. I think it was possible for him to experience a simple emotional response to a building, but it must always have been separated by a membrane of detachment from his complicated intellectual response grounded in engineering, architecture, and economics.

Many years later, I find myself in a remarkably similar situation to what I think my father must have felt that day on the plains of Salisbury. I'm something of an architecture and design groupie. I'm fascinated by the range of effects the design of a building or a streetscape can exert on my feelings and my thoughts; I've traveled the world keen to experience these effects firsthand. Professionally, I'm an experimental psychologist who studies the ways that buildings influence those who use them. I employ a wide-ranging set of scientific tools to get an insider's view of the human response to place. I can tell when the occupants of a building are paying attention (and what they're paying attention to), I know when they are feeling excited, bored, happy, sad, anxious, curious, or intimidated. My mission is to try to figure out the connections between the bricks and mortar that my father measured with such care and the inner workings of the minds of the people who witness them.

I find myself constantly crossing back and forth over a line that separates my simple six-year-old wonder and emotional response to the built world and my critical response as an adult scientist working in the field.

One of my main objectives in this book is to describe both sides of this line to you. Almost everyone in the world experiences built space on a daily basis—in our homes, our workplaces, our institutional buildings, and our places of entertainment and education. We all share at least a vague sense that the way in which such settings are designed exerts an influence on how we think and what we do, and we will often seek out particular settings precisely because we want to experience those influences (think of churches and amusement parks, for instance). But even though we all feel and respond to the design of a building at an emotional level, and even though those feelings influence what we do when we are there, we most often don't have the time or the inclination to dissect our daily responses to place to make sense of them.

Now, perhaps more than ever before, engaged citizens of the world are keen to understand how place works and even to contribute to the work of building better places. In part, this is because we know we stand on the precipice of enormous change. Urbanization, crowding, climate change, and shifting energy balances are challenging us to rethink how we can shape our own environments not only to ensure our survival, but also to guard our mental health. Another part of this new urge to participate in the making of the settings of our lives stems from new tools that are available for us to connect with one another, share ideas, images, and even aspects of our inner mental and physiological states with one another via devices such as smartphones and the Internet.

My own belief is that the key to building better places at all scales is to begin by observing the intricate relationships between our lived experiences and the places that contain them—an enterprise in which everyone can participate—and to continue by bringing to bear the new arsenal of both scientific theory and modern technology to make sense of such relationships. This venture is doubly pressing now because the same kinds of technologies that we can use to study the human response to place—everything from location-aware smartphone applications to embedded sensors that can measure the biometrics of passersby—are also being deployed increasingly in our environment to leverage the traditional tools of design that influence our feelings, wants, needs, and decision-making biases. These technologies are actually redefining everything from public space to the meaning of a wall, and for better or worse, revolutionizing the ways in which our surroundings can affect us. Anyone with serious interest in understanding how place can affect us should pay attention to the many ways that new technologies, admixed with traditional and even ancient methods of building places, can affect our behavior.

The Beginning of Building

The business of designing environments that affect human feeling and action is so ancient that it actually predates any other aspect of human civilization, including written communication, the design of cities and settlements, and even the birth of agriculture, which is traditionally considered to be the seminal event that set into play most of the other forces that shaped modern humanity. The roots of such endeavors lie in southern Turkey, near the city of Urfa, at the ancient ruins of Göbekli Tepe. This structure, more than eleven thousand years old, consists of a series of walls and pillars constructed of stone slabs, some weighing more than ten tons.[1] As architecture, the site represents the oldest building, other than simple dwellings, that we know to have been built by human beings. Indeed, the construction of Göbekli Tepe predates Stonehenge by about as much time as separates the origin of the Stonehenge from the present day. As artifact, Göbekli Tepe is even more important than this. It turns on their heads long-held truths about the origins of architecture. Before Göbekli Tepe, the conventional wisdom was that it was domestication, settlement, and agriculture that spurred the development of architectural practices, and eventually cities. Now it's clear that this story drives the cart before the horse. These stones must have been laid down by hunter–gatherers who lived by killing and eating prey animals rather than by farmers living in settled groups. The walls that have been unearthed here may well be the first ever constructed for a purpose other than to shield the contents of one's possessions and one's family from enemies, the elements, and the prying eyes of neighbors.

Over such a long reach of human history, it is almost impossible to know what purpose the massive columns and walls of Göbekli Tepe might have served for their builders, but the scant evidence of human activity found at the site—the bones of animals and the remains of fireplaces along with the iconography of human figures and large birds, snakes, and carnivorous mammals carved into the columns—suggests that the place served as a kind of religious sanctuary, and most likely, a site of pilgrimage that was visited, modified, built, and rebuilt over a span of hundreds of years. What is clear is that nobody actually lived at Göbekli Tepe. It was a place to visit, perhaps to encourage thought and worship. Possibly, the carvings of the fearsome creatures found there were meant as totems to help manage the fears of the terrible dangers that their designers faced in their day-to-day lives as hunters. It's also possible that, like Stonehenge, Göbekli Tepe was built as a healing place—an indication that one of the earliest human drives that led to building was a response to our awareness of our own finitude and that these early structures represent a nascent struggle against mortality. In some ways, much of the history of architecture, but especially religious architecture, can be seen as a concerted effort to find a way to cheat death—prima facie evidence of our early understanding of the power of the built structure to influence feelings.

Regardless of what can be known about the thinking that lay behind the careful construction of Göbekli Tepe, six thousand years before the invention of the written word, one thing is clear—what happened there may represent the very beginning of what has now become a defining characteristic, perhaps *the* defining characteristic of humanity: we build to change perceptions, and to influence thoughts and feelings; by these means, we attempt to organize human activity, exert power, and in many cases, to make money. We see examples of this everywhere, scattered through the length and breadth of human history.

How Space Moves Us

When I first visited St. Peter's Basilica, I watched as the knees of other visitors buckled at the sight of a gigantic dome dripping with riches and with extraordinary works of art. This very human response is, of course, no accident. Such structures were explicitly designed to change how people felt, to encourage them to reevaluate their relationship with the divine universe, salve their fears with the promise of an afterlife, and hopefully to exert power over their behavior long after they left the site. Indeed, scientific studies have found that exposure to scenes of grandeur, whether they are of a breathtaking natural phenomena like an inky starlit sky or the depths of the Grand Canyon, or a human-built artifact like a cathedral ceiling, can exert a measurable influence on our feelings about ourselves, how we treat others, and even our perceptions of the passage of time.[2]

Our everyday experiences of place are not usually so sublime. When we walk into a courthouse, perhaps just to pay a parking ticket, we can be confronted with high ceilings, ornate decoration, heavy columns or pilasters, all of which help to create a feeling of smallness in the presence of the weight of authority. Again, psychological studies have suggested that the form of such spaces not only affects how we feel, but also our attitudes and our behavior by making us more compliant and ready to conform to a greater and more powerful will.

When we visit a shopping mall or a department store, we may walk in looking for a specific item—perhaps a blender—but we soon find ourselves entering an almost hypnotic state with lowered defenses, diminished reserve, and a heightened inclination to spend money on something we don't need. This state, the holy grail for designers of retail spaces, does not happen by serendipity but by careful design. Ever since we have had disposable income to spend on items that we want but don't need, merchandisers have engaged in a kind of arms race to capture as much of our excess wealth as possible.

When we walk down a wide suburban street with huge setbacks and monotonous tracts of identical cookie-cutter houses, we experience the achingly slow passage of time and a state of boredom not qualitatively different from that experienced by pioneers of sensory deprivation experiments in the 1960s. A walk through a busy, urban street market teeming with colorful wares, the delicious aroma of food, and a hubbub of human activity, on the other hand, can cause our moods to soar. The contrast between our reactions to such spaces can be read easily from our bodies. It is seen in our posture, the patterns of our eye and head movements, and even in our brain activity. Wherever we go, our nervous systems and our minds are massaged by what we experience. Though examples such as the ones I have described may seem so obvious as to be truisms, the subtlety and artistry with which human experience can be influenced by built settings has never been greater than it is now. Not only do designers and architects have at their disposal a wider variety of materials and methods than at any time in the past, but there is also increasing penetration of the guiding principles of the human sciences—sociology, psychology, cognitive science, and neuroscience—into the applied world of design. Powerful new methods in the neurosciences can probe the physical substrate of our mental life with the precision of a surgeon's scalpel. New knowledge of the inner workings of the mind, built on a century of careful experimentation in the cognitive sciences, is bringing us steadily closer to an understanding of the basic building blocks of mental experience that is so sufficiently nuanced we are able to account for and to predict much of our behavior in the chaotic settings of everyday life. At the same time, technologies that permit us to explore the mental and emotional lives of individuals noninvasively and at a distance are accelerating rapidly. We are inundated with new devices that can read our minds via our heartbeats, our breathing rates, our facial expressions, our patterns of eye movements, our sweat glands, and even the patterns of pokes and prods by which we use our mobile phones. Such technologies represent a tremendous boon to researchers in their race to understand exactly how environments at every scale from house interiors to city streetscapes shape our feelings and our behavior, but they also provide unprecedented new leverage in a game that is as at least as old as Göbekli Tepe: the deliberate cooptation of our natural responses to place to shape our behavior.

The New Science of Emotion

For much of our history, common accounts of the make-up of the human mind drew a sharp distinction between the stuff of cognition—perceiving, thinking, reasoning, and deciding—and the woollier territory of feelings, urges, and emotions. We still speak of the dichotomy of "hearts and minds" in much of our everyday discourse, and our literature, films, and television shows are still filled with stories of epic battles between reason and emotion. Our language itself is filled with clues that give away our biases. We speak of "dispassionate" thinking, for example, as if to hold up as the ideal type of reasoning a state of Cartesian withdrawal from urges, hunches, and drives that motivate much of our everyday life. Shakespeare's plays, Jane Austen's novels, and Dostoevsky's greatest works are all filled with stories of the struggles between heart and mind. In a more modern canon, the mythology of *Star Trek*, it makes sense to us that an alien being like Science Officer Spock or the android Commander Data would be capable of perfectly reasoned behavior devoid of the cloud of emotion, and that such behavior could generally be supposed to be adaptive.

Historically, scientific theories have followed the same trend. Outdated theories in the neurosciences have proposed that what marked us as human beings was the ascendance to supremacy of our neocortex, the outer mantle of the brain that housed "higher" functions, where that laden adjective was normally meant to signify pure rationality. Seething beneath the brain's cognitive crown was something sometimes called the "reptilian brain"—the repository of animal urges and instincts that drove us to seek out opportunities for what one wag referred to as the "four f's" of motivated behavior—feeding, fighting, fleeing, and reproduction. In both common discourse and in scientific circles, there was an implicit assumption that these two orbits of our brains—the deeper inside that housed our animal selves and the more evolved outer shell— opposed each other in a perpetual antagonistic push–pull that often left us grasping for reason in a soupy miasma of feeling states we had inherited from our evolutionary forebears. But despite this deep bias in our thinking, modern evidence in neuroscience and psychology suggests a much different relationship between affect and thought. Antonio Damasio's remarkable studies of individuals who have suffered focal lesions to areas of the frontal lobe, once considered the highest reach of rational thought, have demonstrated that such lesions produce deficits in adaptive decision-making and behavior precisely *because* such lesions cut important linkages between our emotional and our cognitive selves. It turns out that those "gut feelings" or, in Damasio's lingo, "somatic markers" that we sometimes use to guide decision-making, and which are more often right than they are wrong, are actually born in our deeper, emotional brains and they constitute important pathways by which we can make sensible goals and plans.[3] Judgment, seemingly supremely rational, is deeply rooted in our states of affect. What we've learned about the important role of emotion in regulating rational behavior from the outcomes of brain injuries has been largely supported by investigations using new methods such as brain imaging and the measurement of brain waves. The areas of our brain that process feelings are widely distributed. They reach all the way from brainstem structures that receive incoming sensations about bodily states, including the states of our hearts, to the higher reaches of the cortex, and they are richly intermingled with structures that produce perception and memory. It is difficult to overestimate the importance of such findings for our overall understanding of how the brain produces adaptive behavior, but such lessons have also not been lost on those with a vested interest in finding ways to influence us. The burgeoning new field of neuroeconomics, for example, is largely founded on the notion that a human being's behavior only follows logical principles so far and that a fully nuanced understanding of how we decide what to do must also take into account our peculiar status as a biological thinking machine, built to survive by means of the principles of natural selection, and subject to biases of various kinds that, though they may not conform to pure logic, have probably encouraged reproductive success. The role of affect in such biases is paramount. At the moment, the application of neuroeconomic principles in the marketplace is an endeavor whose reach exceeds its grasp, but there is little doubt that the gap between theory and application will soon shrink.

These new ideas about the key role of emotion in the guidance of everyday behavior also have great impact on our understanding of our psychogeography—how our surroundings influence us. Though it's far from a new idea to suggest that places influence feelings and that feelings influence decisions, the discoveries that show a profound intermingling of thought and feeling suggest that the extent to which these influences act to change what we do and who we are has been considerably underestimated. But there's more to the new neuroscience that suggests an even closer relationship between our inner nature and the structures and technologies that surround us.

Mirror Neurons, Embodiment, and Technology

Working at the University of Parma in the early part of the 1990s, neurophysiologist Giacomo Rizzolatti discovered a peculiar new kind of neuron in an area of the frontal cortex of the rhesus monkey.[4] Recording from individual neurons using very fine electrodes, Rizzolatti and his team discovered that some cells fired at a high rate when the monkey reached out for a morsel of food, grasped it with its hand, and placed it in its mouth. Such cells, coding for and presumably playing a role in the organization of complex types of actions, are not unusual in primate brains (including ours). What was more notable about Rizzolatti's cells, though, was that they also fired when the monkey watched a video of another monkey completing the same action. Rizzolatti named them "mirror neurons." At the time of their discovery, the importance of these cells seems to have been largely lost on the wider scientific audience. Indeed, Rizzolatti's initial attempt to publish his findings in the prestigious journal *Nature* was met with rejection because it was felt that the discovery would not excite much interest. Over time though, and in concert with other kinds of discoveries about mirror systems in the brain, Rizzolatti's discovery has become recognized as the thin edge of a wedge leading to a dramatically new way of understanding many key problems in psychology, such as our finely tuned abilities to understand and empathize with the behavior of others, but more generally the way that our minds connect to our surroundings. What Rizzolatti's findings suggest is that our brains are built in such a way as to simulate in ourselves the patterns of behavior that we see in others to understand them and, perhaps especially, to feel them. Brain-imaging studies, for example, have shown that when we see someone show an emotion with a facial expression, the brain areas that become active are the same areas that we would use to produce that expression ourselves, as if to suggest that to reach across the boundaries of space to understand what is being felt by another, we must mirror what they show to feel what they feel. In addition, those with brain injuries in areas integrally involved in the expression of emotion also have difficulties with the perception of emotional expressions in others. So the mirror neuron system seems to offer us a way to break through the body wall to make extended contact with other beings and perhaps even with other kinds of objects.

In the rubber hand illusion, participants are placed in close proximity to a model of a human hand and an induction procedure is used to prime them to treat the model as if it were an extension of their own body. The induction procedure consists of their both observing an experimenter gently stroke the hand while at the same time feeling their own hand being stroked. After a short period, participants begin to feel as though the alien hand belongs to them. For example, they will react to the experimenter striking at the rubber hand with a hammer by showing strong physiological responses reminiscent of what would be expected if the hammer was about to pound their own flesh.[5] A similar kind of thing has been observed in what some have called a simulated out-of-body experience in which participants can observe their own bodies from a remote location using an image of themselves projected by a camera to a virtual reality helmet worn by the observer. Using an induction procedure similar to the one used in the rubber hand illusion, it quickly becomes possible to develop the feeling that one is outside one's own body.[6] When I first learned of these experiments, I set up some demonstrations of the phenomenon in my own virtual reality lab. The experiment worked as advertised, creating an eerie and difficult-to-describe sensation of detachment from one's body. It also allowed me the sublime pleasure of watching the chair of my department, an avid participant, lie on the floor wearing a VR helmet while I poked him with a wooden stick.

Similar kinds of phenomena, in which the brain quickly remaps body space to include nearby implements, have been seen in somewhat less bizarre situations. If one is given a long pointer to manipulate objects, for example, brain areas quickly remap in a way suggesting that the tip of the pointer has actually become a part of one's own body.[7] It's not unlikely that the facility that we have for using all kinds of everyday technologies, including the computer mouse, arises from this kind of rapid remapping of the perceived boundaries of the space of the body.

Collectively, what the mirror neuron system and the experiments I have described suggest is that our brains possess powerful and highly plastic mechanisms by which to transcend the barriers between the outer wall of our bodies and any other person or thing that falls within our purview. Such a system not only underpins our ability to use many different kinds of technology ranging from a pencil to a touch screen, but it also suggests that instantiating a bodily state such as a facial expression, even if it is covert, may be the chief means by which we can share feelings with others.

Wonder Woman Poses, Cold Relationships, and Rickety Foundations

In a recent viral Technology, Entertainment, and Design (TED) talk, social psychologist Amy Cuddy described her own research on body language, suggesting that our posture could affect not only our mood, but our body chemistry as well. Her studies showed that participants who were asked to strike "power poses" so that they imitated super-heroes like Wonder Woman performed better in mock job interviews, were more inclined to take risks, and even showed measurable increases in testosterone levels and decreased levels of the stress hormone cortisol after only two minutes of "faking" it. This finding is only one of a recent deluge of experimental reports suggesting that how we move and pose is integrally related to our thoughts, moods, and behavior.[8] One report suggests that peering down at a cellphone makes us less dominant than sitting upright and looking at a larger laptop or tablet screen.[9] Another suggests that holding a warm drink disposes us to be more affiliative and friendly.[10] One study showed that participants seated on a rickety chair were more likely to rate their current romantic relationship as being less stable.[11] Collectively, findings such as these are more than simple laboratory curiosities. They suggest that the linkages between self-produced behaviors of all kinds, from facial expressions to postures and movements, and our psychological states run in both directions. In other words, though conventional thinking might suggest that we smile because we are happy, these findings on embodied states argue that we might also become happy because we smile (and, indeed, this precise effect has been documented in controlled studies).

Given what we now know of the mirror neuron systems first described by Giacomo Rizzolatti, such findings make sense and fit in well with a view of brain organization that suggests the preeminence of body states. We feel because we do. By creating bodily simulations of the effects of emotional states, whether because we have observed them in another or because we have been instructed to do so by an experimenter, we experience both the phenomenology of the state ourselves along with widespread concomitant changes in our physiology, our chemistry, and our hormonal state.

How might such effects spill over into our relationships with built structures? Consider, as an example, the Holocaust Memorial in Berlin. From the outside, this installation looks stark and featureless. It consists of nothing more than a series of black concrete blocks, varying in size and separated from one another by a regular grid of narrow walkways. The heights of the blocks vary and they are set in a ground plane that undulates gently. The power of the monument is largely invisible to the detached observer and only becomes apparent when one sets foot inside the grid and begins to walk around. When I visited the memorial with my wife, we sat on the outside of the grid of blocks for a few minutes trying to puzzle its meaning. Then we set off to explore its inner workings. The walkways were too narrow for us to walk side by side and we soon became separated, catching only brief glimpses of one another as we moved about. As we arrived at the intersections of the walkways, we were able to see all the way to the outside of the monument through long, narrow, empty corridors that skewered us in the gaze of any far-away onlookers standing on the outside of the structure. Our disorientation among blocks that obscured our view of the outside world, the feeling of loss that came from separation from each other, and the occasional visual penetration by the long unoccluded pathways produced a powerful set of feelings: fear, anxiety, sadness, and loneliness. What Peter Eisenman, the designer of the monument had managed to do was to build a structure that resonated with small but potent echoes of many of the feelings that must have been experienced by Jews persecuted during World War II, and he had done this in such a way that the power of the experience could only be had through embodiment. One had to join with the installation, walk through it, and get lost. When one did so, the grief and fear were palpable and convincing.

Although Eisenman's memorial is more of a set piece, designed deliberately to provoke an emotional response from the visitor for the sake of remembrance of the victims whose lost lives it helps us to remember and mourn, we will see in the chapters that lie ahead of us that the power of buildings to elicit such gut responses from us is far from unusual. Either by design or by accident, buildings make us *feel* by making us *do* in ways that are not qualitatively different from the way that mirroring the happy smile of a toddler can make us feel good. These connections are written deeply into our nervous systems in circuits originally designed for us to be able to share experiences with one another and to respond adaptively to the risks and opportunities presented by natural environments.

Smartening Up the Built World

For thousands of years, the conventional wall was a perfectly satisfactory way of influencing human behavior through design. Walls constrain movement and occlude views. They afford privacy and protection. Indeed, John Locke, in his book *Eavesdropping: An Intimate History*, has argued that the wall was designed to protect us from the cognitive load of having to keep track of the activities of strangers as we moved from tiny agrarian settlements to larger villages and, eventually, cities where it was too difficult to keep track of who was doing what to whom.[12] Walls reinforce or perhaps even create social conventions and cultural norms. The invention of dedicated sleeping spaces in homes changed our views about sexuality. The design of traditional Muslim homes and even of streetscapes reified beliefs about gender and generational divisions. Until a century ago, one could have accounted for almost all of the psychological effects of the design of built environments by considering nothing more than the geometry and appearance of the containers of space that were generated by the construction of walls.

But now we have experienced a dramatic change in how we interact with built space that has rendered the carpentered wall, in many important ways, obsolete. The beginnings of this change have old roots: telecommunications technologies like the telephone, radio, and television first made it possible for us to interact with one another, more or less instantaneously, but at a distance, and while out of each other's direct line of gaze. Mass communication provided opportunities for us to share experiences with complete strangers, though such sharing was often anonymous and one-way, as when thousands or even millions of viewers tuned into a popular program or sporting event. But these technologies now seem quaint compared with the world that is emerging—a world in which so many of us walk around carrying smartphones—powerful computers that keep track of our location and movements and allow us to communicate freely with anyone else who carries a similar device. Not only do our phones give us the ability to be connected continuously both with one another and with large repositories of information, but the connections these devices allow are two-way ones. As we tread familiar paths or follow new routes, our devices are shedding information, much of it easily accessible, to the rest of the planet. We carry apps that transmit our location, our activities, and by means of certain kinds of fitness accessories, information about our health. We are traveling beacons of personal data. We are everywhere all at once, broadcasting signals about who we are, how we are feeling, and what we are doing.

It isn't just the marvel of the modern portable phone that fills the airwaves with data related to our movements and thoughts, though. The built world has become increasingly flooded with sensors. Surveillance cameras have been around for years, but now they can be combined with technology that can measure our facial expressions, our patterns of gaze, our heart and breathing rates, and our body temperature. The burgeoning "Internet of Things"[13] joins together every kind of device and structure from the home thermostat to traffic control devices and mass transit ticket systems in a massive electronic skein of information that persistently watches, measures, and adjusts the relationships between people and their everyday settings.

The latest iteration of wearable computing, and the kind that is likely to have the most profound impact of all on our everyday relationships with places, comes in the form of devices that we wear in front of our eyes. Human beings are preponderantly visual animals. Though our other senses play a role in helping us to feel immersed in and to connect with place, it is gaze that most powerfully defines the boundaries of built space. What and whom we can see and how we understand our own visibility to others is the most important determinant of our behavior in the built environment. Because of this, a device like Google Glass is not simply a novel form of portable computer interface, but rather the beginning of a kind of technology that invades that most primal connection. In its current form, Google Glass is not much more than a kind of heads-up display that allows us to receive a steady stream of annotation about our surroundings with nothing more than an upward flick of the eyeballs. But this is really only a short step from a device that might present us with a more complete digital overlay in our field of view that keeps track of our movements and updates what we see accordingly. Such augmented realities have been used in research settings for some time; there are even some rudimentary forms of this way of seeing that are available to users of smartphones. The full penetration of such technologies would, at least for the visual sense, render many of the principles of conventional architecture obsolete. As Joseph Paradiso of MIT's visionary Media Lab describes it, "Everything can become display. Or maybe photons will be painted right onto your retina so it doesn't matter so much what you see. Environments will be some combination of what you physically see and what's virtual."[14]

The power (and perhaps the danger) of such electronic walls is that, unlike walls of masonry that may take years to plan and to complete, walls made of photons can be made and remade in an instant. Not only this, but such walls can be completely personalized. Given the right kind of information, it would be entirely possible for you and me to inhabit the same physical space, but for both of us to see an entirely different kind of vista—one based on our personalities, our preferences, and, perhaps somewhat cynically, our purchasing history. It's already true we each inhabit an individualized world. What we see and how we respond to everyday sensory events has always been conditioned by our unique histories. But when those histories become an open book, available as inputs to providers of technology that can literally put shields before our eyes, then that history will trap us. Rather than becoming an endless source of refreshment and novelty, our worlds run the risk of becoming nothing more than a series of self-reinforcing feedback loops based on something a little bit like our browser history. Everything we see will come to us through the mirror of what has already been seen.

The Way Ahead

Lest I sound like a Luddite, I should say now that I have little desire to return to days when we humans sat around fireplaces in the outdoors, constantly keeping an eye on one another and the scanty possessions we each had arrayed in front of us. I have a healthy appetite for technology, I'm usually an early adopter, and I have no illusions about the many ways in which many different kinds of technologies have made our lives easier and healthier. I can also envision many of the ways in which the use of new technologies that merge the real and virtual worlds of design bring exciting prospects for innovations in responsive environments that will enhance the lives of the elderly, the infirm, and the dispossessed. By way of full disclosure, I should also point out what will become obvious in the pages ahead—that the kinds of developments that I've just described, including mobile data collection, and embedded sensor networks for biometrics and virtual and augmented reality, represent a cornucopia of rich data for scientists who do the kind of research that I do and will describe to you in this book. Put simply, these tools will allow scientists like me to develop a richer and more complete understanding of how the physical surroundings of our lives influence everything that we do.

At the same time, my enthusiasm for this technology, and the possibilities that it holds for transforming how we relate to our surroundings, is tempered by an awareness of the potential for misuse. Increased understanding of the cognitive neurosciences along with vast new opportunities to quickly gather and analyze information about individual behavior in the field gives rise to unprecedented new opportunities to harvest immense amounts of data and to use them to jack into our brains and to completely remake the relationship between us and the world we build for ourselves. Nowhere are these possibilities more acute than in the arena of emotions and feelings—mental states that we now understand to underpin so many of our actions.

Rather than as an alarm bell urging us to step backward to avoid such risks, I think of this book as an attempt to map out the territory that lies immediately before us. As with any great scientific advance, including this one, which seems likely to permeate every aspect of our lives, a better strategy is for us to arm ourselves with knowledge and to hope that wisdom will follow.

CHAPTER 1

THE NATURE IN SPACE

Stories of the Outback

EARLY IN MY CAREER, and as a result of some impulsive decision-making and perhaps a bit of luck (good or bad I wasn't sure at the time) I found myself in the company of a genteel English neuroscientist named Lindsay Aitken, stumbling lost through the Australian outback. We were nervously navigating the tough spinifex grass, terrified by the ticks we'd been told were lurking there, thinking about crocodiles, keeping a wary eye open for snakes, and listening closely for the receding footsteps of our erstwhile guide—an older, seasoned field biologist named John Nelson, who with his forty years of bush experience, vaulted ahead of us with sprightly delight like a jackrabbit sprung from a confining cage. We were out in this thick bush looking for an animal called the Northern Native Cat, a small carnivorous marsupial related to the Tasmanian Devil and unjustly considered by the local farmers in Australia's Northern Territory to be an agricultural pest. Our job was to trap a few of these animals to bring back to our home base in Melbourne, thousands of kilometers to the south.

I confess that at that moment my energies were more devoted to the invention of blasphemous epithets directed toward Nelson and his mischievous glee in having seemingly abandoned us than they were to taking in natural wonders. I can affirm with certainty that any thoughts of biophilia, and the supposedly soothing restorative effects of exposure to natural scenery, were not even on my radar. Just as any other urbanized human might be, I was completely out of my element—frightened, disoriented, and pulsing with adrenaline. Despite my ardent desire to take care of my personal safety, my attention bounced from place to place in a staccato of fearful glances. I suspected the woods were teeming with threats, but I didn't have a clue where to look to guard against them.

When Lindsay and I, red-cheeked and breathless, finally caught up with Nelson, he was standing triumphantly, with a booted foot resting on a stump, and a slight smirk on his face. He pointed toward a gigantic coil of something so big—thicker than my leg—that it took me a moment to process what I was seeing and to realize I was staring at a huge snake: "Python. You can take a photo if you like. He's sleeping. And did you see the crocodile you passed?" Despite his reassurances that he had never really broken contact with us, that he could hear us crashing and stumbling along behind him and knew that we were "fairly" safe, it was some time before either Lindsay or I felt like speaking to him.

What I've described is an extreme example of what happens when hapless modern, urban human beings venture into a truly wild environment, but it highlights an interesting and important fact about the human situation: we have built for ourselves an environment that is so far detached from the surroundings in which our bodies and our minds evolved that most of us, when we are immersed in such wild settings, find that almost all of the mechanisms by which we normally regulate our interactions with place become useless. We don't know how to move, where to go, or even where to look.

Despite our modern state of detachment from the conditions that originally shaped us, most of us still crave contact with nature, though perhaps of a gentler sort than my experience in Australia's Northern Territory. We are innately attracted to elements of places that for our forebears might have made the difference between life and death. Our most expensive real estate often sits on hilltops or on the sides of cliffs facing expanses of water. Even in urban settings we place great stock in privileged views of the natural world. When we visit new cities, we naturally gravitate toward whatever verdant squares and gardens may be on offer. In Vancouver, for example, a city that sits poised gracefully between the Rocky Mountains to the east and the Pacific Ocean to the west, developers of urban real estate are required by law to avoid obstructing vistas of mountain and sea, observing the sacredness of the natural connection.

In the scientific world, and beginning with Texas A & M researcher Roger Ulrich's epochal observation that hospital patients who were able to see some grass and a few trees from their beds recovered more rapidly and required less pain medication than those who could see only bricks and mortar,[1] an avalanche of findings over the past thirty years has bolstered with evidence what most of us feel the truth of in our bones—that nature can soothe, buoy, and restore. If this is true, then it suggests that despite the unfamiliarity of the average city dweller with the grammar and vocabulary of the natural environment, we still possess faint echoes of some deep, primal connection with the kinds of environment that shaped our species. As we shall see, these echoes are written deeply into our bodies and our nervous systems, and they are always at play in shaping our movements through places, our attractions and repulsions from particular locations, our feelings, stress levels, and even the function of our immune systems.

The Biology of Habitat Selection

The question of how any individual animal comes to select a particular habitat is one of the most basic and important ones in biology, and many thousands of research studies have been devoted to it. The ability to select a set of surroundings propitious for foraging, shelter from predators, and the availability of mates is one of the most important determinants of biological success—measured as survival to reproductive age and the production of offspring. Many studies have shown that not only do animals have a remarkable ability to seek out the best available locations for the necessities of life, but that they are able to anticipate how a setting will serve their future needs. For example, the black-throated green warbler, a small songbird that nests in spruce forests of eastern North America, establishes territories preferentially among the red spruce trees of arboreal forests early in the summer months, even though these trees offer less food than the neighboring white spruce trees. Later, though, when nests are built and hungry fledglings need sustenance, it is the red spruce forest that offers the easier pickings. Somehow, the warbler has been guided to settle in locations that will suit its future parenting needs better than its present ones.[2] It is no accident that many studies of habitat selection focus on nesting birds. Construction of a nest requires considerable effort; it is important that the nest site remain stable and safe for the duration of the breeding season, and that it offer the right mix of resources at a time in the future when those resources will be most crucial for the survival of offspring. For many other animals, the problems of habitat selection are slightly simpler. If a location does not offer good forage, the answer may be to simply move along to greener pastures, as herds of elk or caribou might do in following behind the frost line in search of edible moss.

Despite the ample evidence that animals are able to take a broad set of environmental variables into account when selecting habitats, we know remarkably little about the actual mechanisms at play in habitat selection. At the nuts-and-bolts level of perception and movement, what prompts the black-throated warbler to choose the red spruce forest? Why does it *prefer* this type of environment? One reason for the lack of direct knowledge of the precise mechanisms that cause an animal to choose a particular site to dwell, roost, rest, and nest is that it is quite difficult to see into the mind of an animal at large in the field. Laboratory studies have tended to focus on very simple variations in the appearances of environments and to allow the animal to express a preference by means of the amount of time that it spends in an area when given choices. For example, experiments with a type of surgeon-fish known rather delightfully as a Manini (also defined in the *Urban Dictionary*[3] as a "totally fierce . . . rad human being") have shown that in aquarium environments these animals will prefer to spend resting time in shallow areas containing some cover rather than deep or open places.[4] Other experiments with various species of Anolis lizards have shown that they will actually climb poles to survey their surroundings carefully before setting off to find a favorable patch of grass.[5] Though such experiments bring us closer to understanding the mechanisms that guide habitat choices, they still fall a little short of providing a mechanistic understanding of why these animals go where they go.

Human Habitat Selection

Somewhat surprisingly given the wide range of habitats in which human beings can live and thrive, some of the best evidence regarding the mechanisms governing habitat choices comes from experiments with people. Part of the reason for this is that it is somewhat easier to probe the inner mental states of human beings and to measure the extent to which we prefer one type of setting over another using a wide range of different tools from simple self-assessment (aka just ask them!) to measurements of the way in which a person's physiological state varies when presented with a variety of different types of places.

The human preference for particular types of scenes of nature has interested a wide range of specialists since antiquity. Philosophers, artists, geographers, landscape architects, and psychologists have all weighed in. American geographer Jay Appleton synthesized and focused much of this early interest in his magisterial book *The Experience of Landscape.*[6] He took as his launching point biological studies of habitat selection in birds, lizards, and many other kinds of creatures. Appleton described with great significance the German ethologist Niko Tinbergen's argument that the crucial element in habitat selection for an animal was "to see and not be seen." Considered either from the point of view of the hunter or the hunted, the advantages of being able to know what is in the neighborhood while escaping detection of oneself are clear. Appleton, arguing for evolutionary continuity between us humans and other animals, suggested that the same basic principle, reworked in his argument as the concepts of prospect and refuge, might help to account for our aesthetic preferences for particular kinds of natural landscapes. In a way, what Appleton was suggesting was a part of the missing link in biological studies of habitat selection. Perhaps the warblers, Anolis lizards, and Manini of the world gravitate toward the habitats that are adaptive for them because it actually feels good to do so. Implicit in Appleton's argument was that, all modern architectural contrivances notwithstanding, human beings still respond to the faint echo of their natural impulses toward places, even though many of the environmental contingencies associated with those impulses are no longer valid. After all, we're not very likely to meet up with a mortal human enemy or a dangerous predator on a golf course, but the artful design of golf courses that pay homage to the principles of prospect and refuge is one of the reasons why we like so much to be in such set tings even after a morning spent being punished by a tiny white ball. Some have even argued that the perennial popularity of Frank Lloyd Wright's designs, especially his domestic ones, is related to his having had a remarkable, intuitive grasp of the important role that the geometry of prospect and refuge plays in shaping human comfort.[7]

Appleton's description of prospect and refuge helped to galvanize interest in the biological and evolutionary underpinnings of our preferences for certain kinds of scenes in every realm from aesthetics to landscape architecture and interior design; following his work, hundreds of experiments confirmed the importance of this spatial dimension in determining what we like to look at and where we like to be. But it doesn't really take a complicated laboratory experiment to see the truth of Appleton's claim. A quick glance at almost any public space tells the story. In the grand old public squares of Europe, people will most often sit and relax at the edges of the space rather than toward the center. In bars and restaurants where people are able choose their own seats, the tables and chairs around the perimeter of the space will fill long before the central locations do. Even in generic spaces simulated using virtual reality and containing nothing more than blank wall partitions of the kind seen in art galleries, people will show a marked preference for locations that offer the best opportunities for seeing without being seen.[8] Our near universal preference for such locations makes sense to us at a proximate level—everyone knows that we feel more comfortable in such locations—but at the functional level, our preference for spots where we can hunt without being hunted is out of kilter with the contingencies of everyday life. In truth, we are no safer at the edge of a public square than in its center; one could even argue that we can see more of the action from the middle of the setting, which in a people-centered setting like a public square, should make us want to be there. This is the real importance of Appleton's main point—that landscape preference can be understood as a primitive response to a set of risks and benefits that, for the most part, are now irrelevant to our daily lives.

Take Two Trees and Call Me in the Morning

It's easy enough to understand at an intuitive level our preference for certain environmental shapes, but other evidence that highlights the imprint of ancient adaptive preferences on modern behavior is not quite so obvious. For example, in laboratory studies of environmental preference, several researchers have shown a striking preference for scenes that resemble the savannah of Eastern Africa.[9] We like trees that appear in scattered clumps, and we like our trees to have broad, low canopies and wide trunks, much like the Acacia trees found commonly in Africa. Our preference for such settings does not seem to be cultural in origin. Although what I have described sounds very much like the traditional English landscape design so familiar to the Western eye, our preference for such settings does not seem to be cultural in origin. Cross-cultural experiments with participants who live in many different types of environments, including Nigerian rainforests and the Australian desert have shown similar strong preferences for savannah settings.[10] According to the "savannah hypothesis," built on such findings, this suggests that we have an innate preference for the kinds of settings that would have surrounded the East African founder population from which all human beings descended. This preference would have drawn early humans toward savannah settings, which in the face of changes in climate that were occurring, would have conveyed a selective evolutionary advantage on individuals who heeded the call. Like Appleton's prospect and refuge theory, our preferences for savannah settings suggests that we are genetically programmed to prefer to inhabit places that 70,000 years ago would have increased our likelihood of survival. But unlike Appleton's idea, which had to do with the broad geometry of seeing without being seen, our preferences for the appearance and arrangement of trees takes us one step beyond simple spatial considerations and into the realm of color, texture, and form.

These days, it's hard to avoid gushing media accounts of the restorative value of nature. These accounts, though they often focus on the forest rather than the trees, are also based on a considerable body of scientific research that suggests that exposure to any kind of natural imagery—even a lovely landscape painting by John Constable—can exert impressive effects on our bodies and our minds. One of the key pieces of early evidence for this view came from Roger Ulrich's study of recovery rates in hospital patients who had received gall bladder surgery. Ulrich found that patients who could see nature from their windows felt better and recovered more quickly than those who saw only concrete and walls. But the evidence has gone quite a bit further than this, showing that simple views of nature of any form can produce lowered levels of arousal, healthier patterns of cardiac activity, more relaxed patterns of brain activity, and higher scores on a range of psychological tests meant to probe for positive affect. Not only this, but it seems as though when we are immersed in natural scenery, our cognitive apparatus works differently. Measurements of eye movements show that when we are viewing nature we tend to move our eyes differently than when we are viewing urban scenes. Our fixations tend to be briefer; our eyes move from place to place more rapidly. It is as if our attention is flitting pleasantly from one location to another without the strong fixations on small details that are more commonly found when viewing urban settings.[11] These kinds of differences prompted psychologists Stephen and Rachel Kaplan to propose what they called the Attention Restoration Theory. In their book *The Experience of Nature: A Psychological Perspective*,[12] the Kaplans argue that in our normal modern settings, we are called upon to exert a kind of directed attention in which we are required to focus carefully on the tasks of the everyday—anything from completing a routine office task to negotiating a street crossing as a pedestrian. They argue that such tasks are effortful; over the course of time, they drain our cognitive resources. When we are removed from these everyday circumstances and brought into contact with a nature setting—imagine a walk through the woods, for example—we are freed from the need for effortful attention and we are drawn by our fascination with the physical details of our surroundings into a state of effortless, involuntary attention. Our fascination serves to fill the well again so that we may once again enter the fray of civilized life with heightened mood, a relaxed nervous system, and increased ability to focus and attend.

But it isn't just our mood and our ability to think that is affected by exposure to nature. Following landmark experiments by Francine Kuo and William Sullivan[13] conducted in inner city neighborhoods with varying amounts of vegetation, many different studies have now shown that people who live in greener surroundings feel happier and safer. And it turns out that those happy and safe feelings are probably justified, as several well-controlled field studies have also shown that greener neighborhoods tend to have lower rates of incivilities and crime. People who live in greener settings talk to one another more, get to know one another, and enjoy degrees of social cohesion that not only help to protect them from certain kinds of mental pathologies, but also make them less likely to suffer from petty crime. All of these findings suggest that the basic primal response to views of nature, though their origins may have to do with evolutionary factors that may no longer be required to guide judicious habitat selection in human beings, still have significant psychological effects that carry through all the way to crime rates, livability, and happiness in urban environments.

The Mathematics of Nature

Mountains of evidence now suggest that exposure to scenes of nature produces a host of beneficial effects ranging from better health, both mental and physical, to improved neighborhood relationships and happy and safe living arrangements. The work of people like Appleton and the Kaplans suggest that we are hardwired to seek out such settings, presumably because they hearken back to times when being surrounded by the right kind of trees and grasslands enhanced our likelihood of surviving to adulthood and having children, but there is still much to be learned about exactly what in nature produces such effects and what brain pathways might be involved in compelling us to seek green pastures.

One idea has been that there is something about the deep mathematical structure of scenes of nature that our brains are programmed to seek out. Some have suggested that our attraction to a natural scene is related to its fractal properties. To understand what a fractal is, think of a fern frond. The shape of the frond can be viewed at a number of scales—beginning first of all with one entire branch of the plant and gradually descending to the level of its tiny individual "frondlets." But if you look closely at the shapes contained in the plant, you will discover that at each level of scale, from the very large to the tiny, the basic shape that you see is repeated over and over again. This is called self-similarity, or more formally, scale invariance. Scale invariance is seen in many different aspects of nature—think of the branching patterns of trees or even the shapes of coastlines—and it is also seen in human-built artifacts in architecture and art. Indeed, Jackson Pollock's paintings, though they may seem to be nothing more than a random collection of lines and splashes of color, when subjected to mathematical routines, reveal strong underlying fractal properties.[14]

The degree to which a visual scene is fractal in nature can be measured with any one of a number of mathematical routines that yield a number known as the scene's fractal dimension. Understanding exactly how to interpret a given fractal dimension would take us too far into some complicated mathematics, but one way to think about this number is to think first of the dimensionality of simple geometric objects. A line has one dimension. A plane has two dimensions. A sphere has three dimensions. Fractal dimensions for scenes lie between the numbers one and two, suggesting that they are neither quite one- nor two-dimensional geometric objects. In fact, the very name "fractal" is meant to convey this property of having a fractional dimensionality lying somewhere between whole numbers. Though this might seem a bit puzzling to picture, what it really means is that fractal objects defy some of the rules of conventional nonfractal geometry. In his original formulation of fractal dimension, the Polish mathematician Benoit Mandelbrot considered how one might go about measuring the length of a jagged coastline using a measuring stick. Because it contains a vast number of detailed curves and angles, the measured length of the coastline will depend on the length of the stick. As the stick becomes shorter and shorter, the length of the coastline will seem to become longer and longer. Fractal dimension describes the relationship between the length of the measuring stick and the measured length of the coastline. If the coastline happened to be a perfect straight line, its fractal dimension would be 1, so not really a fractal at all.

Using mathematical tools that aren't much different from unleashing a range of measuring sticks of different sizes on an image, it's possible to arrive at a number that characterizes the fractal dimension of the image. When these tools are applied to scenes of nature, the measured fractal dimension often lands at a value somewhere between 1.3 and 1.5. What's interesting about this is that psychological studies, some using a variety of scenes of nature and others using more artificial images (fractal art, abstract patterns, and even Jackson Pollock paintings), have shown that people prefer to look at images that have approximately the same range of fractal dimension as that found in nature. This correspondence between the fractal properties of images and our preference for them, and even in some cases our physiological responses to such images, which can resemble the restorative response to natural scenes, has given rise to the idea that the way that the brain actually recognizes nature is by means of this mathematical property.[15]

The idea of an account of our attraction to natural scenes based on the mathematics of fractals has a certain appeal. For one thing, fractal dimension is a property that is easily quantified (though of course there are lots of squabbles among scientists about the exact way one should go about doing this). In addition, there is a pretty elegance to a theory of attraction to natural scenes that is based on mathematics and that could also be used to predict our attraction to any kind of scene, whether or not it contains objects of nature. On the other hand, despite many years of intensive investigation of the physiological properties of brain areas that process information about the visual world, there has never been a single report of anything like a "fractal detector." So although the fractal idea is appealing, in the absence of a serious theory of how the brain goes about detecting fractal patterns, the fractal idea lacks biological plausibility.

In response to this serious issue, one of my graduate students, Deltcho Valtchanov, set out in a somewhat different direction—still focusing on mathematical properties of images that might be used to predict our attraction to them, but looking for some kind of property that is known to be of interest to the neural pathways that process visual information. Valtchanov didn't have to look far. Another way of characterizing images has to do with their spatial properties. The easiest way to understand this is to first realize that any image is made up of a set of lines and contours of differing width and contrast. What I really mean by this is that most images of real things (as opposed to weird images conjured in visual perception laboratories) contain big, blobby contours (think of what a severely out-of-focus photograph looks like—such an image is showing only low-frequency contours), and finely detailed contours (think of an exquisitely detailed Rembrandt etching that contains lots of closely spaced fine lines). Every image will contain a broad range of these different types of contours ranging from the very tiny to the very blobby and large; it is the mix and proportion of such contours that ultimately gives any image its final appearance. In fact, the mathematical underpinnings of this statement have proven that we can conjure any image using a judicious combination of entirely abstract patterns of light and dark stripes of varying thickness and contrast.

As it turns out, our visual systems are replete with neural circuitry that is designed to work out these kinds of details. Cells at all levels of the visual system from the retina to the upper reaches of visual cortex are specifically tuned to be looking for a particular size resolution of contours, and the mix of such specifically tuned cells can vary from one area to another. This makes perfect sense in that different types of information are buried in images at different levels of detail, and different brain areas are charged with sorting out different kinds of information.

Valtchanov asked whether the complement of contour types found in an image—called the power spectrum of the image—might have something to do with our patterns of preference. He explored this question by collecting large numbers of different kinds of images and manipulating their power spectra, using image-processing software like Adobe Photoshop. He then presented these sets of images to participants in our laboratory and asked them to rate their preferences. Astonishingly, he found that the power spectrum of an image was a strong predictor of preference, even when the images were degraded to a point where they were barely recognizable. Even more interesting than this, he found that the power spectrum of an image could predict how much we liked to look at it, even when comparing sets of images that didn't contain any natural scenery at all, but only depictions of urban streetscapes.[16]

Valtchanov's theory relating contours to human preference has some of the same advantages as the fractal theory—it is grounded in the mathematical properties of images and so can be used to make strong predictions about what we like. But Valtchanov's idea has the further advantage that the mathematical properties he uses to make his predictions do have strong biological plausibility. We've known since the 1960s that lots of cells in the visual system are supremely concerned with contour thickness (in the scientific literature, referred to as spatial frequency for reasons that would take us a bit too far off the beaten path). Not only this, but Valtchanov's notion that spatial frequency may be the missing link relating scene preference to basic mathematical properties of visual scenes fits very well with other findings in scene perception, suggesting that the power spectrum of an image is a key factor in our ability to recognize the basic properties of different kinds of scenes very quickly. Experiments in scene perception suggest that we can extract the gist of a scene—whether it is a cluttered forest, an open beach, or a busy streetscape, for example, in an incredibly short time—something in the neighborhood of about twenty milliseconds (much less time than it takes for an eye blink, in other words).[17] This kind of rapid extraction of gist is undergirded with visual processing mechanisms related to image power spectra.

Where in the Brain Does Nature Preference Reside?

Whereabouts in the brain the most important visual processing related to scene preference might occur is still something of an open question, but recent brain-imaging studies have shown that there is an area in the temporal lobe of the brain, known as the parahippocampal place area or the PPA, that is nestled among a cluster of other areas that are involved in complicated processing of visual information about objects. The PPA is very interested in scenes, collections of objects laid out in a natural arrangement, as they would occur in the real world. Cells in these areas have some interesting properties. For one thing, they seem to be very responsive to the property of enclosure, perhaps pointing directly to the neural basis of our preference for Appleton's refuge. But more importantly for our present purposes, the PPA is happiest with scenes that contain strong proportions of spatial frequencies in the range that Valtchanov has shown is most predictive of human preference.

For the icing on the cake, there is one other interesting property of the PPA that makes it a strong candidate as the headquarters of a circuit that controls our emotional responses to scenes, and perhaps the long-sought-after missing link in our understanding of the biological mechanisms underlying human habitat selection. This area of the brain is unusually rich in opiate receptors. These neurochemical receptors, long associated with brain mechanisms that control our perception of pain and naturally occurring analgesic effects such as the "runner's high," are also heavily represented in the brain's reward pathways. At the neural level, one part of what happens when something makes us feel good, whether it be a good meal, a romp in the sack with a lover, or the ingestion of a drug like heroin, is the activation of opiate receptors in the brain. The presence of such receptors in a part of the brain that is ostensibly involved in processing visual scene information is compelling evidence indeed that we are on the right track to discover exactly which pathways are involved in generating pleasurable responses to place.[18]

Simulating Nature

When we began our laboratory investigations into the psychological effects of nature, our ambitious goal was to try to pinpoint the underlying basis of our attraction to natural scenes. Although I think we have made some progress on this, we had other aspirations as well. In the face of an explosion of new technology for the presentation of scenes on large screens and unprecedented new opportunities for making those scenes interactive in some way, we imagined being able to bottle the magic somehow and to find ways to produce strong restorative effects without any real nature in the picture at all. In some of our early experiments, long before we started presenting participants with weird, degraded images of nature and of cities, we built a virtual reality environment designed to give people a compelling and immersive experience of nature. Wearing headsets that contained small screens and that responded to every step and head turn of the viewer, we were able to place people into virtual rainforests, jungles, and beaches, replete with the sights, sounds, and in some cases, even the smells of nature that were so convincing that some of our participants forgot about the real space where they were standing—a room with spare furnishings, many computers, and lots of wires. In these experiments, we deliberately produced high levels of stress in our participants by asking them to recollect unpleasant events from their lives or by asking them to do difficult mental arithmetic while listening to industrial noise. When we moved them from these stressful conditions to a pleasant virtual forest, their psychological readings swung hard and fast toward positive values in less than ten minutes. And scenes of nature were far more effective than control scenes depicting urban settings, so it wasn't just the release from stress that we were seeing, nor the simple novelty of being able to play with some cool machines. In fact, the effects that we saw were larger than those reported by others who had done similar experiments but who had "restored" participants by immersing them in *real* nature.[19] I find myself mixed about this finding. On the one hand, our ability to reproduce the restorative effect using pixels on a screen provided us with a powerful tool that we could use to advance our understanding of the effect. But on the other hand, I worried (and still do) about the potential of such findings to somehow suggest to us that real natural settings, especially in cities, could be supplanted by the wizardry of technology. If we don't need the real thing to reap the psychological benefits of nature, then why not dispense with it altogether and get on with the job of city-building using massive multicolored screens as building façades and piping in the sounds of waterfalls and chirping birds?

I can certainly imagine circumstances where the ability to reproduce nature and its effects on individual psychology would be tremendously advantageous. Imagine, for example, the benefits that might be experienced by someone who was otherwise unable to venture into a natural setting because of some kind of infirmity. An elderly person confined to a wheelchair, a shut-in, someone with an advanced disease state that precludes being able to wander freely down a forest trail might still be helped to enjoy some of the restorative benefits of nature by the use of technology.

There are other special circumstances that might also be perfectly suited to technological interventions simulating nature exposure. One of my colleagues, for example, has helped to pioneer the use of virtual reality scenes of nature as a supplemental form of analgesia during dental surgery. Several experiments have shown the potency of immersion in a techno-forest for ameliorating both the anxiety and the pain that sometimes accompanies such medical procedures.[20] A company in the United States, the Sky Factory, has begun to market artificial skylights that contain either photographic images of nature or high-resolution movie players that present dynamic nature scenes. These devices have been installed in hospitals, chemotherapy suites, doctor's offices, and other locations where patients might obtain a measurable boost from experiencing a dose of nature while undergoing a painful or stress-inducing procedure.

Alternatively, imagine what it might be like to live in a superurbanized setting surrounded by a natural environment that, though beautiful, might pose serious risks to casual visitors. In Malaysia, for example, residents of large, dense cities like Kuala Lumpur live surrounded by lush jungle that could afford tremendously enriching opportunities for communion with nature, but visitors must also contend with poisonous reptiles and insects along with a handful of powerful predators who might interrupt their gentle walks in much the same way that my own experience in Australia's Northern Territory was affected by my lack of proper understanding of the risks of life in the bush.

Finally, without proposing to actually replace natural scenes and trees in urban settings with screen-based simulacra, it's possible to imagine that understanding the principles involved in the healing effects of natural imagery might provide avenues by which we can supplement rather than replace urban nature so as to enhance the opportunities for restorative experiences in dense, urban settings or building interiors where it would otherwise be difficult or perhaps even impossible to include true natural elements.

Psychologist and author Peter Kahn, in his thoughtful book *Technological Nature: Adaptation and the Future of Human Life,*[21] has discussed some of these ideas in the context of experiments that he has conducted to investigate the prospects and limitations of replacing true nature with a variety of technological innovations. In one study, for example, Kahn compared the effects of providing research participants with a view of a garden through a glass window with a view of exactly the same scene captured using a webcam and presented on a plasma screen attached to the wall of the testing room in the same position as the window. Surprisingly, Kahn reported that the screen view did not produce any physiological signs of the restorative response. In a follow-up study, Kahn showed that when technological devices (again, wall-mounted screens) were provided to office workers who had no windows, the outcomes were more positive. Participants who were provided with such screens reported that they enjoyed using them to view natural scenes; they felt that the devices enhanced the quality of the time they spent in their offices and their productivity.

In the final analysis, these experiments suggest that, when we have no alternative, we can find psychological succor in technological simulations of nature, but when a real window is available, a screen-based substitute has only a minor effect on us. It's hard to know exactly what to make of these findings. One possibility is that the screens, based on the technology of the day, were lacking some important properties. Kahn talks about the property of parallax, in which the view that we see through a window shifts slightly as we move around in front of it. Screen-based views currently lack parallax, so it isn't hard for us to discern that they aren't real. It's also possible that participants, because they knew that they were seeing mere images of nature rather than real scenes, were somehow discounting the value and meaning of the scene and thereby showing less of a deep psychological response to what they were shown. In the experiments with interior offices where occupants showed more positive responses to the displays, the explanation might be that such workers, because they were used to being in a situation where a view of nature was a frank impossibility, had a different kind of baseline response and so were more sensitive to the inclusion of a screen view in their otherwise somewhat bleak surroundings.

Regardless of the ultimate explanation for the effects that Kahn reports, peculiar in light of my argument that at least some of the biophilic response is based on the purely visual properties of nature scenes and also somewhat at odds with much research that shows some degree of restoration for images, videos and even abstract fractal designs, his research serves as a warning that any proposals to replace natural scenes with technological substitutes should be approached with circumspection. Such substitutions may produce some of the same effects as immersion in nature itself, but perhaps only in special circumstances when there are no ready alternatives.

Civilizing Attention

There is a much deeper question to be answered about the relationship between technology and nature that reaches far beyond considerations of the technical details of display technology. To see the outlines of this question, we need to reach further into the past to consider how we built a modern world that requires us to engage in a daily struggle to tap limited cognitive resources. If the natural condition—the one that lowers our stress levels, produces psychological fascination, healthy and joyous patterns of attention shifting—was so good for us, why did we sacrifice it? What did we trade for it? In his book on technology and nature, Kahn begins by describing the arduous lives of Kalahari Bush People living in the traditional way. Enduring harsh conditions with extremes of climate and unimaginably heavy exertion to chase prey or lug edible roots over long distances, these people never saw a square, carpentered corner, or a paved road or sidewalk. In what sense did they enjoy advantages over modern life? According to Kahn, and based on Elizabeth Thomas's somewhat idyllic accounts of life among the Bush People as the daughter of pioneering anthropologists Laurence and Lorna Marshall,[22] their way of life represents one of the most successful cultures to have ever existed. They lived "free and wild" with nature, cohabiting peacefully and intimately with an environment that offered them everything they needed to survive as a culture for about thirty-five thousand years. What changed? Though a complete answer to this question would lead us far beyond our scope, some of the answers have to do with shifting patterns of settlement brought about by changes in climate and the development of agriculture. Unlike small, nomadic groups of Bush People, pastoral settlers growing crops for food soon found themselves living in larger settlements and investing labor in infrastructure that made nomadism impractical. These larger, fixed groupings of people brought about pressures for new social arrangements, trade, political hierarchies, and according to Lewis Mumford in his encyclopedic *The City in History*,[23] a "citadel" mentality in which occupants of a large settlement began to place themselves into defensive opposition against the wildness of nature. Over the course of centuries, this opposition brought with it the walls, ramparts, tools, and weapons—in other words, technology—that allowed the urban environment to flourish as a full antinomy to the immersion in the wild that characterized earlier nomadic hunting cultures like that of the Bush People. But although it's not hard to see the traces of this gradual move away from nature in the development of larger cities filled with strangers, human conflict, and most importantly, the development of physical structures that furthered our separation from the conditions in which early humans came to be, there is even more to this story and it is of a much more recent vintage than the development of the world's first cities.

There are many separate strands to the story, some involving changing views of the organization of our minds, but others involving industrialization and the mechanization of mass production. Jonathan Crary has brought many of these strands together with admirable clarity in his book *Suspensions of Perception: Attention, Spectacle and Modern Culture*.[24] He first describes important changes that were taking place in the scientific realm with the birth of scientific psychology and changing views of how the senses were organized. Studies of both the psychology of perception and the physiology of the sense organs had begun to reveal that there was a more ephemeral relationship between the external world and the internal mental representation of that world than had been commonly assumed. Although it was not particularly new territory in the philosophical realm to draw a distinction between the realm of the senses and an ultimately unknowable external reality, hard data from newly emerging laboratories studying the psychology of perception were revealing empirical truths that were beginning to put to rest a notion sometimes called naïve realism—that we sense what we sense simply because that's what's there—and to replace it with the idea that human perceivers are active observers who *construct* a sensible interpretation of whatever their sense organs might be telling them. This important shift in the way that the role of the perceiver was understood had consequences that reached far beyond the cloistered laboratories of early psychologists. Most importantly, it meant that human beings willfully brought their perceived world into being by deliberately drawing together the facts of sensation into a coherent story—often doing this by focusing on some aspects of their senses and ignoring others—put simply, by paying *attention*.

At the same time that work in psychology was beginning to revolutionize ideas about how we understood the world given to our senses, other kinds of changes in economics, mostly related to industrialization and mass production, were changing the way in which workers were viewed. As employees on factory floors were becoming increasingly treated as commodities, so were their perceptual systems, and especially their ability to use these systems to complete routine tasks. In other words, the human ability to pay attention was also becoming commoditized. Indeed, though we may have mistakenly mythologized Thomas Edison as the inventor of the light bulb, his real genius lay in his understanding the vital connection between the organization of the human mind and the principles of mass production. Just as Edison understood the value of a nimble and plentiful power grid to large-scale industry, he could not have failed to notice that the proper application of scientific principles to the worker himself would yield productive advantages. No less important than this was his understanding of the power of media to shape our habits of consumption. Long before Marshall McLuhan[25] revolutionized our understanding of the power of media to shape thought, Edison's invention of the "kinetoscope," a forerunner of modern motion-picture technology, along with his contributions to other kinds of communication technology such as the stock ticker, revealed the early development of an understanding of how the presentation of text and image could be used to shape our willful acts of attention. By harnessing the power of such technologies to capture thoughts, influence life narratives, and create strong appetites for consumption, Edison contributed to the beginnings of a trend whose effects have accelerated wildly for the past two centuries.

Today, even in the face of a mass of evidence that nature is good for our minds, we still place the highest premium of all on our abilities to maintain task-oriented laser-beam focus on activities that will contribute to our productivity. We treat our refreshing interludes in natural spaces as periods of rest from a "real life" centered on production and consumption. Our educational system, especially at the primary levels where minds are the most malleable, is almost entirely based on the precept that the goal of formal education is to produce an individual who is capable of sitting still and *paying attention* to a single focus of activity in the classroom. Children who have difficulty doing so are often singled out, pathologized, or even treated with drugs that change the functions of their brains in such a way as to promote tightly focused selective attention. Indeed, the very structure of classrooms at every level from the kindergarten to the university lecture hall is designed to reinforce the virtue of the kind of focal, effortful, directed attention that becomes depleted quickly.

Screen-based technologies of every kind, from the gigantic electronic billboards of the world's Times Squares to workstations, laptop computers, tablets, and mobile phones, represent the natural continuation of technologies designed to attract and hold captive what has become humanity's most precious cognitive resource—our attention. But before there were any screens, architectural technologies as simple as walls achieved the same ends. By hiding or revealing particular elements of the world, walls also serve to focus and direct our attention.

Considered in this light, one can view much of the history of modern environmental design, beginning with the arrangement of bricks, mortar, plaster, and windows and ending with the development of electronic screens that function as a very potent type of artificial window on the world, as a systematic assault on our natural way of seeing and being in the world. Our natural habits of attention, and the ones that we still seek out for small moments of refreshment in our harried lives, have been replaced with a perpetual state of sharply focused, selective attention that both helps to create desires and equips us to satisfy them, but ultimately leaves us mentally depleted. The technologies of attention have pulled us irresistibly away from the kind of lifestyle enjoyed by pretechnological societies like the Kalahari Bush People, integrated with the natural order. Instead, we have become a set of neurological machines honed by our environments to be optimal producers and consumers. Indeed, there is some irony in the fact that we now see the primordial wild environment that brought us into being as a kind of temporary escape valve that we can use to rescue ourselves from the cognitive consequences of a long preoccupation with consumerism—the invention and satisfaction of our ever more complicated material desires.

Given this radical transformation of what it means to be a modern human being, perhaps the most remarkable fact of all is the clear evidence for the continued influence of the imprint of our early origins on our current feelings, our preferences, and our behavior. Although I think that few of us, myself certainly included, would dream of replacing the comforts of our modern built settings with a brutally hard life carved from the wilderness, there is no doubt that we still crave the vistas and the natural geometries that would have enhanced our prospects for survival in settings that we walked away from thousands of years ago. One can see the hallmarks of these preferences in almost every aspect of our behavior, from where we choose to walk and sit, what we like to look at, and how we try to arrange our lives, alternating as much as possible between the powerful forces of technologies that shape attention and the restorative effects of natural settings—of both the authentic and the simulated kind. More than any other single factor, our cravings for nature underlie the psychogeographic structure of our lives.

CHAPTER 2

PLACES OF AFFECTION

Living Sculptures, Loving Buildings

I N THE QUIET CENTER of this small fern forest, I could feel my heart slow and my muscles relax. Disconnected thoughts that had raced through my mind after a frenetic drive along a busy freeway gave way, replaced by a feeling of calm and contentment. I had retreated to the quiet center of myself. I felt, in the term used by scientists who study our responses to natural environments, as though I was "away" from my regular life. Time slowed down. My eyes began to flit effortlessly and pleasurably from place to place as I immersed myself in the setting and I became fascinated by it.

I reached out to touch a frond that was at eye level. At first it curled up a little, but then it reached out to touch me. This is when things began to get a little weird. This was no ordinary forest. If I pushed, it pushed back. If I flinched, it responded by moving toward me with curiosity. This forest seemed to know that I was there and I could easily imagine that it knew a little about how I felt. My initial absorption in my surroundings began to give way to something less familiar. My feelings shifted in the direction of uncertainty, surprise, perhaps even a frisson of threat. When I walk through forests, I'm accustomed to being surrounded by life of all kinds—singing birds, chirping insects, and the natural sway of vegetation in the wind. What was happening in this forest was different. It wouldn't be unusual to notice that one's presence in a forest had been noticed—birds and insects might stop their activities, sensing a human threat—but here I felt as though I was the very center of attention. The forest responded with obvious interest and intent to every move that I made. I felt exposed. This forest seemed to know me.

The small forest I was standing inside, remarkably, was entirely artificial. It was in the living room of a beautiful old house in the leafy suburbs of Toronto that doubled as the workshop of the architect Philip Beesley, the visionary creator who had designed this grove of plastic ferns with a 3D printer, a large collection of simple microprocessors and sensors, and some spools of a special kind of wire called resistance wire that expands and contracts in response to electric current. This mass of delicate filigreed acrylic petals was a small working sample of a series of major installations that Beesley has set up at a number of international exhibitions, including the Venice Biennale in 2010, where hundreds of thousands of visitors have wandered through several different giant versions of Beesley's so-called *Hylozoic Soil* series, experiencing the same kinds of strange feelings that now enveloped me in his workshop. The response to his work has been nothing short of sensational. Evoking states of intimacy and connection in visitors, Beesley says that his intention is to gestate feelings of sympathy and care, but to build them "out into a sense of exchanges in space, where the boundaries of what and who I am, the differences between me and an animal and a rock, become quite blurred."[1]

I had met Beesley for the first time several years before my visit to his workshop. I was involved in a fund-seeking venture for a project looking at the use of new kinds of technologies for measuring feelings and behavior in community health clinics. I had heard that Beesley's architectural practice had been involved in the design of several such clinics, so on the urging of another architect on the team, I asked him if he would take part. When I convened our first meeting, I sat in a conference room with a handful of professionals from different disciplines ready to discuss strategy. Beesley arrived late, sweeping into the room with a broad, beaming smile and infectious energy and enthusiasm, but looking slightly off balance, like a man who had far too much going on. As most of us were strangers to one another, I suggested we begin the meeting with a quick round of self-introductions. Others in the group gave the usual boilerplate messages, describing their discipline, their qualifications, and how they saw themselves fitting into the project's mandate. When Philip's turn came, he told us that he wasn't quite sure how he might fit in with the rest of us, but his main interest these days was in the design of special kinds of sculptures that lived at the edge between life and nonlife, generating a strange admixture of attraction and revulsion in observers who would become enmeshed in and ultimately digested by the creations. There were a few seconds of silence at the table—unusual for a gathering of talking heads—as we got our first inkling that this was a man who would not allow us to plod along with pedestrian ideas about what a building could or should be. Beesley was in another universe, it seemed, with a vision of a world far off in the future, but also firmly rooted in our ancient past.

A quick glance at Beesley's curriculum vitae gives some clues as to how an architect graduating from the University of Toronto in the mid-1980s went from a conventional practice consisting of the design of residential housing, student centers, health centers, and restaurants to a set of interests he describes like this: "Emotion, romanticism, and twentieth-century spiritualism as alternate qualities in Modernism; alterity and dissociation; chthonian and expanded definitions of space; the archaic."[2]

As he describes it, a turning point in his life was a project he carried out with the support of a prestigious *Prix de Rome in Architecture* award in 1995–1996. Working with archeologist Nicola Terrenato on the ancient ruins of the Palatine Hill in Rome, Beesley's own role in the dig was to try to reconstruct the circumstances of an ancient sacrificial burial of an infant, deep in the foundations of the structure. The burial, taking place in the eighth century BC at the Porta Mugonia, one of the three ancient gates of the original city, was an example of a common ritual of the times in which the children of the first families inhabiting a city were sacrificed at its borders to define a threshold separating the wild outside from the urban interior. Beesley's experience here, carefully dissecting the tomb of a baby, examining its painstaking construction, thinking about its meaning, set in motion a lifelong preoccupation with thresholds between living and nonliving, the capture of vital forces in meshes of human construction and eventually to the world of geo-textiles—fabrics that act on soil—and from there to *Hylozoic Soil*—fabrications that are neither living nor dead, that respond to living beings and take on some of the most intimate properties of that life: empathy and caring.

Beesley's work is characterized by a series of remarkable leaps of imagination backed up by careful scholarship, deep thought, and an ability to draw long arcs of connection between seemingly separate realms of discourse and understanding (one of his most recent projects involves designing clothing with Iris van Herpen, Lady Gaga's dress designer). These abilities shine through not only in his creative work but in his everyday persona as well. His language, gestures, and expressions lead his listeners along a merry chase of fascinating ideas extending from high theory to fundamental architectural practice. When Beesley visited my virtual reality lab, I placed him into an immersive simulation representing a modest piece of domestic architecture of which I was quite proud. Others who had seen the work stood still, looked around, made a few tentative steps toward objects, sometimes reached out to try to touch things and asked a few questions. Beesley dove into the simulation with gusto and challenged it from the inside. To the alarm of the students who controlled the machines and managed the wires connecting the pieces of gear together, Beesley was soon running from place to place, crawling along the floor to get underneath things, lying on his back on the floor looking up at ceilings, consuming my simulation with happy, childlike curiosity while those around him scurried about trying to keep wires and computers in check.

Beesley's artistic works are both moving and thought-provoking, but they seem far-flung from the bread-and-butter design of buildings like schools, banks, offices, or homes. Like the haute couture world of fashion, in which we see models walking runways in outfits that most of us would not be caught dead wearing in the street, Beesley's responsive sculptures can be thought of as a set of signposts to the future: the bleeding edge of what design in a wired world has in store for us, and one of the main subjects of this book. *Hylozoic Soil* provides a high-impact object lesson in the extent to which a *thing* can develop a two-way emotional relationship with a human being. The somewhat unsettling collection of feelings that is conjured by a walk through Beesley's empathic sculptures contains some of the thin strands of the tapestry of human emotion that we know as "love" and hallow above all other states of being.

Everything Seems Alive

More words have been written about the state of love than of any other human capability, feeling, or emotion, and definitions of the word extend into antiquity. Scientists have gathered data, extracted blood samples, and measured brain waves, all in an effort to reduce to numbers what it means to love. But most of this research has centered on love of the interpersonal kind—the love that entices us into long-term pair bonds, children, minivans, and mortgages. It's very common for us to express love for an object or thing, but except for the case of paraphilias like shoe fetishes, we normally mean something quite different when we say to a friend, "I love that dress!" than when we declare eternal love for our romantic partner. Yet there are those who maintain a romantic relationship with built structures. Eija Riita-Eklaaf, a resident of northern Sweden, has declared her love for the Berlin Wall (or what's left of it), and has actually carried out a marriage ceremony of sorts, renaming herself "Wall Walther." Erika Eiffel (née Erika LeBrie) famously married the Eiffel Tower in 2007, declaring her infatuation with its long sinuous curves. U.S. native Amy Wolfe married a thrill ride at Pennsylvania's Knoebels Amusement Resort. Though it's easy enough to write off these so-called Objectum-Sexuals as nothing more than crackpots, it's hard for most of us to deny that we have felt strong attractions to certain kinds of objects, shapes, or features. For reasons that I could not possibly explain, I once had a strong fondness for a beautiful red can opener that was given to me as a gift before I moved into my first home. I hadn't realized how much I liked this kitchen implement until it was finally taken from me by rust. Canned beans have never tasted the same.

The allure of the beautiful curves of a Mazda Miata does not come out of nowhere. Our primal responses to the geometry of a sports car are written deeply into our nervous systems. Not only have researchers shown a preference for curves over sharp angles—a preference that begins in infancy long before our experiences have taught us the perils of sharp-edged objects like knives and scissors—but these preferences also seem to be related to the properties of nerve cells in areas of our visual cortex involved in object recognition. Put simply, we have many more cortical cells devoted to the analysis of the nuances of a curved surface than of a sharply angled one. These cells are part of a very rapid neural processing system for forming first impressions and evaluating threats. Even our first impressions of strangers whom we meet are based in part on analysis of simple facial features related to shape. Without being aware of it, we form preferences for certain types of faces within less than thirty-nine milliseconds of their appearance. That's about a *twentieth* of the time that it takes for the average human heart to beat *once*.[3]

Beesley's kinetic sculptures go far beyond employing a simple rubric of connecting feelings and shapes, however. Although the overall appearance of the sculpture does have an organicity resembling a natural forest, a set of features long known by environmental psychologists to promote relaxation and even healing, what is crucial to the envelope of feeling that connects object and viewer has more to do with movement and interactivity than geometry.

In 1944, Fritz Heider, a perceptual psychologist at Smith College, and one of his students, Marianne Simmel, published research that showed the prepotent tendency of human beings to attribute higher-order states of mind, including intentionality—a sense of purpose—to simple colored geometric objects presented in a short film. Participants in these experiments saw nothing more than a pair of triangles and a circle moving around on the screen. When asked to describe what they saw, participants described the unfolding events in very human terms, attributing cognitive and emotional states to the objects. One viewer of the video described one of the triangles as an "aggressive bully," for instance, and many viewers speculated about the possibility of a love-triangle between the shapes. These much-cited experiments helped to lay the foundation for a psychology of "theory of mind," which supposes that we are predisposed to attribute the behavior of all kinds of objects to very human inner feelings and thoughts. More recent work has suggested that the development of the capacity to use theory of mind to account for simple perceptual phenomena is something that begins at a very young age. Even infants show some of the effects described by Heider and Simmel.[4]

In a related vein, the Belgian psychologist Albert Michotte reported experiments in 1947 showing what he called the "launching effect." In Michotte's study, participants were shown a display that was considerably simpler than the one used by Heider and Simmel. Michotte's film showed a red dot moving toward a green dot on a screen. When the red dot made contact with the green dot, the green dot moved away. Michotte's participants found it impossible to describe what they had seen without invoking the idea of causality. The red dot has somehow caused the movement of the green dot. Much subsequent research has shown that the launching effect is not only remarkably robust, but that it is essentially impossible for us to perceive the simple display in any way other than as showing a causal relationship between the events we see on the screen, even though what is actually being shown is nothing more than the movement of a pair of dots. And, like Heider's findings, Michotte's demonstrations of perceptual causality have been shown to hold in very young infants.[5]

The experiments of Michotte and Heider collectively suggest that human beings are hardwired to see simple moving objects as sentient beings capable of complex emotions like love and jealousy. Findings such as these go against the common-sense (and incorrect) view that, when we look at a scene, we first struggle to identify and categorize all of the objects that it contains and only then go on to puzzle out what is going on. Even our superquick first pass through the scene, carried out in much less than a heartbeat, includes these automatic inferences of cognition, feelings, and intentions. In terms of the evolutionary pressures that helped to shape our nervous system, it's not hard to understand why this might be so. In trying to make sense of the world, the human brain faces a prodigious problem: there is far too much information out there for us to plod carefully through all of the data that are available to us in the average scene. Not only this, but the brain, as a biological computer or "meat machine" as it has been called, is an astonishingly slow processor. Compared to the artificial computing devices that we build—even relatively simple computers such as the ones that help our cars run smoothly or allow our iPods to play songs for us—the brain works at achingly slow speeds. To compensate for this sluggishness and still deliver the on-time performance that will allow us to dodge oncoming predators (which could be anything from saber-toothed tigers to hurtling Buicks), our brains employ a veritable Swiss Army knife of tips, tricks, and short cuts. One of the most important of these types of conjuring tricks is that our brains are designed to anticipate what kinds of things *might* be "out there" based on what kinds of things are *normally* out there. Some of this anticipation is learned on the basis of past experience, but learning itself can be slow and painstaking; for many types of situations there are no second chances. We simply don't have the luxury of learning the consequences of *not* dodging an imminent threat. Instead, much of the assumption-making that our nervous system does is built in from the start and is so automatic that even if we try to ignore it, we can't. So we see Heider's moving triangles as a pair of fighters dueling over a mate and we see Michotte's colored circles as if they were a pair of billiard balls bouncing off one another, even though they are really just simple shapes on a screen.

So now, returning to Beesley's *Hylozoic Soil* installations, it's a little bit easier to understand why a visitor would feel immersed in an unsettling miasma of feelings when confronted with a sea of swaying acrylic ferns. The magic that is taking place has less to do with the ferns being able to penetrate deeply into our limbic brains and much more to do with the fact that they are tapping into a mechanism that has evolved to help us understand real-world situations with blinding speed. This mechanism may also help to explain certain features of pathologies like Objectum-Sexuality and even hoarding disorder, where sufferers often feel deep and emotional attachments to ordinary, household objects. One victim of hoarding disorder, for example, related her experience when trying to dispose of some empty plastic containers in her recycling bin. After washing them carefully in the sink and then throwing them away, she recounted that she was unable to stop thinking about the fact that the containers might be feeling uncomfortably damp because of their recent wash, so she went back to find them, removed the lids and dried them carefully to allay her concern.[6] Such bizarre thought patterns may represent a more extreme version of a universal human tendency to hold animistic beliefs produced by a nervous system that is wired to make lightning speed judgments and decisions. This same inbuilt neural tendency to animism also probably holds the key to ancient sacrificial rituals like those that Beesley studied in which we have joined living human beings with the fresh earth of a new city.

Loving Homes

If, as Beesley has demonstrated, it is so easy for us to strike up strong facsimiles of interpersonal relationships with a piece of sculpture, then what happens to us in our homes? If any kind of space is built for intimacy, it is the one to which we retreat at the end of a hard day in the world of work for rest, succor, and protection.

The idea that our homes may be sentient is one that enjoys a long and rich literary history, and often not with happy outcomes. In Edgar Allan Poe's chilling story *The Fall of the House of Usher*, the author takes steady aim at the notion that the setting for the story, a spooky Gothic mansion, is really one of its main characters. The narrator, describing in rich detail the gloomy setting of the house interior, explains that the house conjured a frightening state of mind, but "wondered to find how unfamiliar were the fancies which ordinary images were stirring up."[7] Just as Philip Beesley's aim was to design a set of "exchanges in space" that blurred the divisions between observer and environment, the narrator of Poe's story describes a harrowing dynamic in which the players and the setting feed off one another in their march to an inevitable and horrifying conclusion. Much more recently, the successful

American cable television series *American Horror Story* played on the same themes, albeit with a heaping dose of graphic gore.

But when they're not scaring the wits out of us, our home spaces are much more customarily associated with a set of positive values. We look to them for privacy, acceptance, comfort, and intimacy. This connection is made with graphic and beautiful simplicity in the construction of traditional homes in West Africa's Mali, for example, where the layout of the home is designed explicitly in the shape of the female figure, and the central living space is, literally, in the womb.[8]

Western-style homes, for complicated reasons that have more to do with economics than with the psychology of design, have evolved well beyond the traditional forms of vernacular architecture in which home-dwellers were free to design their own abodes using whatever local materials they could find. But we can still identify many of the ways in which the design of domestic spaces exerts a steady impact on how we behave at home, and especially, how we interact with our housemates. In his best-selling book *Home*, Witold Rybczynski describes the evolution of the home from a simple one-roomed affair with movable furniture and essentially no privacy to the stately homes of the wealthy; he argues that this progression has gone hand-in-hand with our discovery of the importance of comfort, a concept that was completely lacking in early domestic spaces.[9] Others have described more explicitly some of the relationships between the arrangements of home spaces and different aspects of our personal and social lives. For example, the innovation of dedicated sleeping spaces for married couples was an important milestone in our changing views of sexuality and privacy. But it wasn't just that the construction of these new sleeping arrangements satisfied a desire for intimacy between partners. Rather, it was a two-way street. The very existence of separate bedrooms actually promoted the virtues of a sex life conducted behind closed doors, and helped to change our understanding of the relationship between parents and their children. Similarly, the development of dedicated rooms for the preparation of food created a private domain for whoever was in charge of preparing the meals, and helped to reinforce the notion of specialized domestic roles for husband, wife, and child within the home. Indeed, Peter Ward, a sociologist of space and author of *A History of Domestic Spaces*, has gone so far as to suggest that the arrangement of more complex Western home spaces with their affordances for privacy, territory, and a room of one's own contributed to the Western trend to value the individual over the group.[10] The possibility of constructing a life apart from other people, even other members of our family, encouraged us to place a high value on our independence and autonomy. There is no end to the ways, both subtle and fairly obvious, in which the changing arrangements of our homes reinforced new patterns of behavior and thought about our place in the grand scheme of things. Hermann Muthesius, a German diplomat who worked in London early in the twentieth century and author of the magnificent two-volume history of English domestic architecture titled *Das Englisch Haus* went so far as to suppose that one of the reasons for the economic success of the British in comparison with Germany at that time had something to do with the arrangements of their houses. As he saw it, English houses were designed to promote a comfortable and informal division between intimate spaces and the more public areas of the home designed for entertaining guests. In contrast, he argued that German homes were designed in such a way that visitors experienced something like a brassy Las Vegas attraction in which each act had to surpass the previous one as guests were led from one ostentatious room to the next in a rigidly prescribed procession of spaces.[11]

Experiments with Virtual Homespaces

If it's really true that the appearance and the layout of our domestic spaces influences our feelings, and that the right kind of house can make us fall in love with them, then it ought to be possible to measure these kinds of interactions scientifically. Until recently, the right kinds of tools to make these measurements didn't exist. For much of the past century, psychological experiments most commonly took place inside stark laboratories where participants sat motionless in front of a stimulus while investigators asked them questions or occasionally poked and prodded them with scientific instruments to measure muscle tone, heart rate, eye movements, and sometimes brain waves. Now, with the advent of much more sophisticated technologies that can keep track of moving observers, we have available to us more sophisticated methods for measuring our responses to space.

One of the most powerful approaches of this kind is based on the methods of immersive virtual reality. With these methods, participants are shown images of places, which can either be presented on small displays contained in a helmet or even projected onto the walls of a room. The real magic of virtual reality is that what observers see on the display is updated according to how they move. Sophisticated sensors measure every movement of eye, head, and body, and the images presented on the screens change in synchrony with these movements. Using this kind of motion tracking, it is possible to put the observers right inside a computer model of any kind of design in such a way that they see it in its full three-dimensional splendor and they feel as if they have been transported to an alternative reality. Although the viewers seldom lose the sense that they are in a simulated space, they act in many ways as if they have left the real world and entered whatever creation the programmer has designed for them. So, for example, anyone who is afraid of heights will become measurably anxious when placed into an open-air virtual elevator. These kinds of powerful visualization methods are slowly becoming stock-in-trade for enlightened architects. Before constructing an actual building, they can throw together a facsimile made out of pixels that they can show to clients and troubleshoot for errors. But researchers interested in how we interact with spaces have also begun to take advantage of this technology; environmental psychologists have used such methods to study our responses to places. Virtual reality approaches are likely to increase considerably as the cost of a good simulation system has dropped precipitously in recent years.

In my laboratory at the University of Waterloo, we decided to take advantage of this emergent technology to study how people responded to different kinds of domestic spaces. We built computer models of three different home designs. The Jacobs house was built by Frank

Lloyd Wright in 1936. The interior of the house was a modestly sized L-shaped design replete with warm natural materials, and as free as possible of ornamentation, bric-a-brac, and trim, reflecting Wright's belief that a home should celebrate the freedom and autonomy of the individuals who live in it. We also modeled a house designed by Sarah Susanka, a prominent U.S. architect and author of the well-known *Not So Big* series of books, which celebrate small but extremely functional and comfortable designs.[12] The third house was a fairly typical modern North American suburban tract house. Participants in the experiment were wired with virtual reality helmets and they were invited to treat the experience as if they were shopping for a home and had come to visit these three particular models. They were followed carefully through each of the homes, free to travel where they wanted to, and accompanied by an interviewer who asked them a series of questions during their walk. Not only were we interested in hearing people's impressions of the houses, but we also wanted to keep careful track of where they walked and where they looked as they moved through the virtual spaces. With the experiment under way, it became obvious to us very quickly that participants soon forgot that they were only exploring a computer model, and they began to treat the simulation as if it were the real thing. One participant, for example, noted that when she walked past a long row of picture windows in the Jacobs House, that she could feel the warmth of the sun on her hands—this despite the fact that the "sun" in our model was a complete fiction and exuded no heat at all. Other participants would carefully duck underneath imaginary kitchen cabinets to peer into corners of the design. When we asked participants to indicate their favorite locations in the homes, most of them gravitated to the largest open spaces in the center of the living areas, telling us that they liked locations from which they could see what was going on. When we asked them where they would place a treasured family heirloom, people chose a variety of different locations—some wanted the heirloom on display in the most prominent part of the house, whereas others hid objects away in secret corners of back bedrooms.

When we tracked the route that people followed during their explorations of the houses, we were surprised to find that there were certain rooms that were not entered at all. Most prominently, the large formal living room in the suburban house was peered into from both of its entry points, but it was not explored at all. Strangely, many participants told us that they liked this room, but apparently not well enough to want to be in it! This finding perhaps mirrors the complaints of many who own such homes that formal spaces are not well-used and are a waste of valuable ground-floor real-estate.

Our biggest surprise, though, was that when we asked people which of the three houses they would be most likely to purchase, there was a strong tendency to indicate the suburban house despite the fact that most people did not say that they liked this house the most! Participants marveled over the creative use of space in the Susanka house, its affordances for both privacy and socializing, and its practicality. Others lauded the generous use of natural materials in the Wright house and its unique and inviting living space centered on a large hearth. Yet most people said they'd be unlikely to buy either of these houses.

The most likely reason for this disconnect between what we see and feel in a house and what we want to have in our own lives has to do with the past experiences of the participants. They told us that though they found the designer homes interesting and attractive, they gravitated toward the kinds of houses that they were most likely to be able to find in the current market. So to some extent one could see this sad disconnect as a failure of imagination on the part of the participants. We want what we want because it is all we feel that we can have. But as I pored over the numbers I wondered whether there might be some more profound influences at play. How does our history of home-dwelling, the things that have happened to us in the places we have lived and the memories that have been captured by the rooms where we've lived, influence our feelings toward domestic spaces? When we fall in love with a house, what exactly is it that we arc falling in love with? The spaces and surfaces that we experience as we enter a house for the first time may be only one part of the story.

Real-estate agents will often tell clients that, when they have found the house that they want to buy, they will know immediately because they will feel it. Presumably, this assertion is made on the basis of their experiences with many home-buyers. And as artificial as our virtual reality experience might have been, we actually saw some shades of this kind of response in some of our own participants, especially as they entered the Susanka model. There was an immediate and palpable sense of engagement with the space. They slowed down and looked around more, taking their time to relish the experience. Just by listening to their comments and watching their movements and without even looking at the numbers we collected during their visit, we knew too. So what is it that we know? Where does that dizzy love-at-first-sight feeling come from when we visit a house for the first time?

Gaston Bachelard, a French philosopher and poet, gives a detailed account of the phenomenology of lived space in his book *The Poetics of Space*. There, he describes the manner in which our early experiences with homes can shape our lives. The home, he says, is above all else a kind of container for our daydreams. It is in our first home that our routines of thinking and memory are first formed, and we can never really sever our connections between those early experiences and our later actions. He says

> ... we are very surprised, when we return to the old house, after an odyssey of many years, to find that the most delicate gestures, the earliest gestures suddenly come alive, are still faultless. In short, the house we were born in has engraved within us the hierarchy of the various functions of inhabiting. We are the diagram of the functions of inhabiting that particular house, and all the other houses are but variations on a fundamental theme.[13]

These unconscious connections between our earliest dwellings and our current living circumstances are probably common. We've known since ancient times that there is a special linkage between our life experiences and memories and the places in which those experiences took place. One of the oldest memory tricks in the world, the method of loci, first described by Cicero in his *De Oratore*[14] written in 55 BC, prescribes an explicit strategy of linking memories with places to improve one's memory of them. In modern times, thousands of experiments in psychology and neuroscience have substantiated Cicero's claims.

For example, Gabriel Radvansky, a psychologist at the University of Notre Dame, has discovered a striking dependency between our memories for the objects in our lives and the environments in which we find them. Radvansky and his team set up simple memory experiments in which participants were asked to carry objects from one room to another, leave others behind, and meet yet other objects in the rooms to which they moved. He showed that as we pass through doorways, we seem to move from one "event horizon" to another, making it more difficult for us to remember the objects that we left in the room behind us. Careful control experiments showed that it was not simply an effect of the passage of time or the physical movement from one place to another that produced the effect. For example, even carrying out a complicated route in the same large room did not lead to the erosion of memory. There was something special about passage through a doorway. The effect held regardless of whether the participants walked physically from place to place or navigated through a virtual environment; even small depictions of three-dimensional environments on computer screens where participants moved using a mouse produced the effect.[15] Radvansky's studies have placed some of the phenomenological observations of philosophers like Gaston Bachelard under the microscope of cognitive science and have paved the way for a more detailed scrutiny of the interactions that take place between experience, memory, and the rooms in our homes.

The idea that our mental representations of home consist of an amalgam of the seen and the remembered has powerful implications for a psychologically based science of design. Perhaps most importantly, it suggests that a designer who wants to make a home that can be loved cannot simply tabulate a list of physical features that have been shown to be pleasing to the human perceptual apparatus. The designer must understand the client's history, the kinds of dwellings that they have known so far, and the kinds of things that happened to them in these places of memory.

But the things and places of our lives have an influence that extends far beyond lists and catalogs of remembered events. The feelings we experience are inextricably linked with such memories and they can exert a considerable influence on our attachments to home spaces. Several years ago, I was the guest on a call-in radio show on the subject of our use and understanding of space and place. I won't soon forget the story of a man who had called in to recount his own story in which he found himself strangely drawn to purchase a house after a brief first glimpse. He came to realize that the basis of his attraction was that the house resembled his childhood home. Unfortunately, the man had experienced some traumatic childhood events in this early home of his, and he soon discovered that the recognition of his early house, far from triggering warm reflections of happy family life huddled around the hearth, began to bring back these more frightening images for him. He told me he had become involved in a desperate bid to remodel the house so as to disguise it from himself, to somehow build a chimeric space that preserved the normal functions of the house, but put him at arm's length from those early unpleasant memories. We may store our memories in our houses, but we take them with us when we leave. If we are lucky, they can form the backbone of a happy adult life, but our early memories, if they are not so benign, can spring their mental coils at unexpected moments, leaving us with unsettled feelings reminiscent of those felt by the poor victims of Poe's stories.

Art therapists know that our mental representations of our homes may contain clues to an unhappy past. Françoise Minkowska, a Polish psychiatrist and colleague of Hermann Rorschach of the famous eponymous test, studied house drawings made by Jewish children who were victims of persecution during the Nazi regime. She described the drawings of such children as being tall, thin, relatively featureless, and forbidding. More recent studies have shown certain hallmark features of the house drawings of children who are the victims of abuse. Their houses lack doors and are often filled with sharp contours, and oddly, a profusion of hearts. There are often storm clouds above the house and falling rain, features seldom seen in the house drawings of children who are not abused. Even as children, or perhaps especially so, we see our homes as surrogates for our own psychologies and experiences, and that is how they are represented in our imagination.[16]

Pioneer of psychoanalysis Carl Jung describes his attempt to design such a home for himself in "The Tower," a chapter in his autobiography describing the tower's construction that has become required reading for students of architecture. In this work, he gives a fascinating account of the interplay between the construction of the home, his developing theories of the organization of the human mind, and developments in his own life. Starting from his understanding of a home space as representing an intimate facsimile of a mother's love, he saw the original round tower as a form of oversized hearth representing a womblike environment much like the West African homes that are shaped like human figures. Over time, the round tower called out to him for several additions, built over the course of more than a decade, until it had the feeling of completeness, with each of the main elements of the tower representing one of the structural rungs of his encompassing theory of the psyche.[17]

There is much more to the story of Jung's tower that is barely mentioned in his autobiography, but which can be gleaned from his voluminous notebooks. Jung had a lifelong habit of working through difficult problems with his hands. After his well-known and emotionally charged split with Sigmund Freud, one of his closest friends and his most important intellectual mentor, Jung decided that the path to survival lay in a thorough exploration of his earliest memories of childhood. An important part of these memories constituted the construction of different types of miniature castles—a practice that he took up again at Böllingen, the site of the tower, along the shores of the lake that would be his home for much of his life. The tower that he finally built for himself as a home represented a life-sized version of these childhood recollections, but he also included within the structures of the tower much that could be considered homage to the important events, ideas, and people of his life. He says

> Words and paper did not seem real enough to me. To put my fantasies on solid footing, something more was needed. I had to achieve a kind of representation in stone of my inner-most thoughts and of the knowledge I had acquired. Put another way, I had to make a confession of faith in stone That was the beginning of the tower, the house I built for myself at Böllingen.[18]

Home Spaces for the Rest of Us

Few of us have the resources or the opportunity to build a dwelling that reflects our inner psychologies from scratch. It's more likely that like participants in my virtual reality studies of domestic spaces, we will only be able to select from among a few different types of homes that are available at the time that we're in the hunt for a home. As we've seen, we may find ourselves gravitating to the familiar, even when our brains and our bodies might be sending us faint signals suggesting that we are selecting against our own preferences simply by force of habit. In a few lucky cases, we may strike upon homes that are in harmony with our ancient memories.

But considered on a global basis, the kinds of experiments and obser-vations about home spaces that I have been describing do not reflect the relationship that many of us have to the spaces where we live. Home spaces can run the gamut from a high-rise apartment building in Shang-hai to a spot on the pavement under a flyover in downtown Mumbai. Indeed, even in a superdeveloped country like the United States, a large study conducted by the Pew Research Foundation showed that roughly one-quarter of Americans don't consider where they are currently living to be "home."[19] In short, most of us don't really get to choose our homes. Rather, they are thrust upon us.

In the absence of opportunities to select a home that is a perfect fit to our personal aesthetic or our experiences, how do we develop attachments to our home spaces? In many cases, such attachments may be rooted in the possessions that we take with us as we move from place to place. Immigrants moving from one continent to another, for example, may take little else with them than the clothes on their backs, but they usually make room for religious artifacts such as the family Bible or shrine, albums of photographs, and a few other items that they must use to try to draw connections between their ancestral homes and their new lives. For them, the only visible daily connections with their earlier homes are made through these simple sets of possessions. They use them not only to anchor themselves in their new abode, but also to exert a form of control over the appearance of a home space that might afford little other opportunity for personalization.

Even in the most spartan of surroundings, we struggle hard to personalize our homes to exert control. A walk through Dharavi, Mumbai's largest slum settlement, reveals a wild array of colors and vernacular designs using whatever materials can be scraped together to produce a vibrant community teeming with activity and commerce (including a bustling economy generating over 600 million U.S. dollars annually). Conditions in Dharavi are immensely challenging in many ways—the supply of water and electricity is sporadic and there is very limited access to proper sanitation—but there is strong evidence of a powerful struggle by its residents to control their own surroundings as much as they can.

In an entirely different context, U.S. architect Oscar Newman, in his classic treatise *Defensible Space,* emphasized the importance of feelings of ownership and control over communal areas in high-density living spaces like St. Louis's notorious Pruitt-Igoe social housing complex.[20] Newman described design principles that he argued could increase the safety and livability of such housing projects; he contended that it was the failure of Pruitt-Igoe to enact these principles that led to its demise. The main aim of his design toolkit was to find ways to encourage feelings of place attachment in environments where economic challenges acted against any desires that the tenants might have had to personalize and take ownership of their living environment, much as residents of Dharavi struggle to own their space. Although the reasons for the demise of Pruitt-Igoe have undergone reconsideration recently, and probably had as much to do with chronic underfunding as they did with design, Newman's principles have merit and are still put into play to try to reduce crime in dense and poor neighborhoods.

Human crowding in tiny homes packed tightly together can cause entirely new challenges for our definitions of home. In our spacious Western homes, whether we have found true love for them or not, we normally think of them as places of privacy and refuge. While conducting psychogeographic research in Mumbai with the BMW-Guggenheim Laboratory, a traveling think-tank focusing on urban issues in the world's major cities, I discovered a curious anomaly. When I took people into sparsely populated public places like museum parking lots or churchyards, they showed strong evidence of the kind of relaxation response that one might normally see in a quiet and private environment like the home or a beautiful green park. Using sensors that measured physiological arousal, I could actually see their bodies calm down in response to these empty places. In the Western context, an empty public place might be considered a failure: Most of our planning efforts in such places focus on the problem of bringing such spaces to life. These measurements did not surprise my assistant, Mahesh, who described his own domestic arrangement in Dharavi where he lived in a single small room with his wife, two children, his parents, and his two brothers. Mahesh said that for him, the way to find peace, a space to share an intimate moment with a friend, or to enjoy a moment of solitude was to *leave* the hubbub of his home space and find a quiet place somewhere in the city. Mahesh's anecdotal observation was strongly reinforced by research that was parallel to mine, conducted by Aisha Dasgupta of the BMW-Guggenheim Lab in collaboration with PUKAR, a local research collective studying urban issues in Mumbai. Although 54 percent of respondents in a survey designed by this group still described their home as their most private space, they also indicated a tendency to seek solitude in public places, though they decried the relative scarcity and inaccessibility of safe public spaces, especially for women.[21]

Considered on a planetary basis, the variety of different types of arrangements of space that we call "home" is so broad that it defies easy classification, yet the psychological principles that come into play are somewhat easier to digest. Some design features have near-universal appeal. We are attracted to certain shapes and colors, especially those that contain some of the same elements that are found in nature, and this can include things like the views from our windows. We gravitate to the shapes of spaces that provide some privacy and feelings of security. Our feelings of attraction to our home spaces are also dictated by our individual histories. Our early experiences, and the places in which those experiences took place shape our adult preferences by either attracting or repelling us to certain kinds of arrangements of spaces, depending on the valence of those early life experiences. Finally, our love of our home is dependent on our feelings of control of the space—the extent to which we have been able to shape it to our own individual psychology using anything from our prized family heirlooms to simple elements of décor like posters, paint, and wallpaper. When we lose the struggle to achieve that kind of control, our love of home can die on the vine.

Home's Future

But what of the future? Are we forever doomed by the exigencies of the economics of the cookie-cutter house or condo to be offered homes in generic structures that may contain a few appealing design flourishes to tickle our senses, but which otherwise have no more connection to our inner workings than as if we were simple two-dimensional images of ourselves in the glossy pages of real-estate brochures? Or worse: as we are compelled to occupy smaller spaces in the crowded and expensive cities of the future, will we have to abandon completely the idea of an individualized domestic space tailored to our own preferences, experiences, and inner psychology?

If forward-thinking architects, taking inspiration from Beesley's kinetic creations, have their way, then the future of domestic designs may look entirely different. What if your home, rather than being the four mute walls of a box that contains your domestic routines, could be a more active player? What if your home could help you to fall in love with it by *returning* that love? This is the future promise of responsive design.

The idea that a building can have senses and can respond to events that take place within its walls by adapting is not a new one. In a way, you can think of the thermostat that controls your heating and air-conditioning systems as being a sort of responsive system. The thermostat takes an input—the temperature setting that you select— that represents a simple form of desire and through the magic of a feedback loop, it operates the complicated mechanical systems that work to meet your needs. There are many such simple control systems in homes, ranging from fire and intruder detection systems to auto-mated devices that can control lighting and media systems for enter-tainment, but these kinds of systems, decentralized and very much under the control of a human user, are different in kind from the idea of a home whose sensing and responding systems are tied together and under the control of an intelligent agent that works to adapt the house continually to the occupant's needs.

Nicholas Negroponte, cybernetics guru and founder of MIT's well-known Media Lab, was the first to suggest that architecture could be conceived as a kind of computing machine that could respond to and interact with its users. Writing in the 1970s, Negroponte foresaw the promise of combining computer power with building materials in such a way that a structure could respond intelligently to events that took place in and around it.[22] Most work so far in this field has focused on finding ways to enhance environmental sustainability in buildings by design-ing features that minimize their carbon footprint. The North House, for example, a design that originated in the University of Waterloo's School of Architecture and was spearheaded by Beesley, was a response to the challenge of building a zero carbon-footprint house in northern cli-mates. The North House accomplishes this by means of a set of sensors that can respond not only to the weather outside, but also to the internal environment and the positions, movements, and activities of its occu-pants. On a larger scale, responsive envelopes for major buildings have been developed that minimize energy costs by responding to environmental events. The Design Hub at Melbourne's Royal Institute of Technology is composed of thousands of polished disks that rotate to follow the sun, lowering the energy costs of the building, and one day serving as a power plant using an array of photovoltaic cells. Similarly, Chicago architect Tristran D'Estree Sterk designs shape-shifting buildings whose organically curved and pleasing outer envelopes use the principles of what Buckminster Fuller called "tensegrity"' to alter their very form in response to readouts from sensors. So far, these sensors are designed to measure things like air temperature and sunshine, with the main goal of producing a pleasant internal atmosphere with the minimum of energy expenditure. Although such structures certainly take advantage of modern sensing gear and materials to produce accommodations that make for greener buildings, they are really only a small step beyond the simple feedback mechanism of a thermostat on the furnace of a house. These buildings may know certain things about their occupants, but they do not feel them. Now, even some living room gaming consoles contain simple sensors that can measure our heart rates, our stress levels, our facial expressions, our eye movements, our breathing rates, and our brain waves, so it is conceptually within reach for a building that is studded with such sensors to have available to it more information about our physiological and mental states than might be available to a close friend sitting with us in our living room.

Some of the more recent projects at MIT's Media Lab point in this direction. One such project, the CityHome, an invention of the Changing Places Group headed by Kent Larson, is composed of a number of separate modules that allow users to configure apartment-sized spaces to suit their needs through movable walls embedded with computing devices that can keep track of the physiological state of the user as well as their behavioral history. These so-called living laboratories can be configured to collect enormous amounts of biometric data about their occupants, and certainly enough information to construct a reasonable theory as to their state of mind and body. One could almost think of the CityHome experience as being something like having a live-in butler who is ever responsive to the needs of his charges, anticipating and making a more commodious setting for their every move, but—and this will probably be important—ready to disappear into the woodwork at a handclap.

It isn't just the temperature and lighting that can be customized for the comfort of a home's occupants, but also almost every aspect of the appearance of its inner surfaces. Daniel Vogel, a computer scientist and artist in Waterloo's Cheriton School of Computer Science, has worked with giant-sized interactive display panels that can respond to hand gestures or changes in posture. In an age of cheap and fast motion tracking technology, like Microsoft's Kinect for example, such displays are within easy reach of hobbyists, and very simple for professionals to put together (I recently visited a virtual art gallery with full-body motion tracking and good voice command recognition that had been designed by a computer scientist acquaintance of mine after about a week of work). As Vogel says, "you don't want it to become like Times Square in your home,"[23] but nevertheless it is now an attainable goal to have ultrathin display materials that might show the weather outside on your bedroom ceiling, the news on your bathroom mirror, and a panoramic window into a nearby park on your living room wall. Given our predilection for certain kinds of images (of nature, for example) and the known psychological effects of certain kinds of colors, patterns, and images, it is entirely within reach for a sentient home that knows how you feel to modify its appearance accordingly. Feeling unwell? Your home might dim the lights and soothe you with the lapping waves of a beach at sunset. Looking for inspiration to achieve a looming deadline at work? Your home brightens the lights, shows you an inspiring scene of busy human activity in a teeming city square, and puts on the coffee. Combine this level of interactivity in your home with a psyche that we know is predisposed to see animated life in the simplest of gadgets, and it is very easy to imagine that we will soon be able to enjoy a new kind of relationship with our homes: one in which the fabric of the building is a living being with its own mind, personality, and a set of feelings for its occupants built over a long history of friendship.

It's easy to think of the advantages of such a home setting, especially for those of us with physical or mental ailments. Imagine a home that knows when you are beginning to feel depressed and responds accordingly by engaging you in an encouraging conversation or by suggesting an activity that could brighten your mood, perhaps even alerting your flesh-and-blood friends if things begin to look dire.

A new smartphone app that is just hitting the marketplace now can measure your mood based on your patterns of usage of social networks and even the manner in which you engage with its touch-screen, and it can respond by alerting a health care provider. A sentient home, by learning your habits and having a window into your physiology, could be even more proactive in helping to care for you. The possibilities for increasingly sophisticated home awareness systems for people who have special needs have become a strong focus of interest, especially given our overtaxed health care systems and an emphasis on finding ways to allow the elderly or ailing to live independently at home rather than in institutions. But even for those of us who have no such special needs, a home that can feel and respond to us has its appeal, and could be quite subtle. As we've seen, the human tendency to see life and complicated sets of emotions in simple geometric shapes or lines is very powerful, and could be parlayed into a fully fledged kind of emotional prosthesis that could help to accentuate positive emotions and to blunt negative ones.

The advantages of the sentient home are easy enough to imagine, but in a culture with dystopian visions of automated overseers like HAL 9000 (from Kubrick's movie *2001: A Space Odyssey*) or MU-TH-R 182 (from Ridley Scott's movie *Alien*), it's possible to imagine a down side. As such modern sci-fi fables have taught us, intelligent computer interfaces can make mistakes, misinterpret their instructions, or be hacked for nefarious purposes. But these are the kinds of risks that accompany the use of any kind of technology upon which our lives and well-being may depend. Perhaps a deeper concern has to do with the risks of allowing our technology to supplant functions that might constitute important parts of the essence of what it is to be human. In the same way that some have argued that a world full of GPS signals has eroded our native ability to find our way around, and that powerful search engines have given us lazy memories, a house that buffers us from the real contingencies of life might similarly blunt our exposure to certain types of realities.

The emergence of any kind of technology causes a fracture between our lives and what is real. We welcome it because it frees our time and mental capacity for other kinds of pursuits, but we may not necessarily be aware of all that we have given up to obtain this freedom. If our lives really consist of the collection of a set of routines that focus us on the world, on one another, and on our understanding of ourselves, then what might it mean when the full bloom of technology is allowed to invade the very materials used to construct our last bastion of refuge—the most intimate and private space that most of us will ever know? Though the symbology of our homes may connect with life in the womb, do we really want to return to it?

As new design concepts in domestic spaces continue to incorporate the staggering potential of information technology to transform the definition of home, we will continue to grapple with questions like these, which may still seem in some ways to be like science fiction. But as the wired world makes further inroads into our everyday lives, it will focus more debate on what might be seen as a kind of crisis of authenticity. With the acceleration of cheap and easy technology in several advancing fields—sensor design, display devices, tensegretic structures, virtual reality, and 3D fabrication—we will experience the new sharp edge of what the French poet and philosopher Paul Valéry called "the conquest of ubiquity." This idea was further elaborated by the German essayist Walter Benjamin in his influential essay *The Work of Art in the Age of Mechanical Reproduction* proposing that accurate mass reproduction of artifacts will require new ways of thinking about what the word *real* might mean, while other developments in technology and social networking cause us to struggle with concepts such as privacy, autonomy, and authority.[24] A transformation is coming in which the very setting of our life story will be less of a passive stage on which we act and more of an active participant in the process. As with all epochal changes, we can simply allow events to envelop us and take whatever comes, or we can meet these challenges with debate, discussion, open-minded experimentation, and hopefully some optimism.

German philosopher Martin Heidegger worked for much of his life in a small cabin in the Black Forest, not far from where he was born. His most important works, including the transformative *Being and Time*, were written in the tiny hut that he shared with his wife and his two sons. He may have seen, in his seclusion from the rest of the world, among the hills and mountain peaks of his home, a metaphor for his own views on the difficulties of doing philosophy, which he likened to "speaking from mountain top to mountain top." He eschewed the high manner and dress typical of German academics of the time and maintained a rustic appearance and style of speech that was very much in keeping with his surroundings. His famously difficult language, including many new words that he invented, or compounded from other words to make new terms, also seems to resonate with the confusing complexity of the hilly paths he walked around his home. Even one of his most important works, a collection of essays known in English simply as *Basic Writings*, was titled *Holzwege* by Heidegger, which translates literally as *Woodpaths*, but in Germany is understood to signify a confusing path through woods that makes linear progress from one place to another very difficult. Woodcutters speak of *holzwege* as paths that one must follow to find wood, perhaps as fuel for a hearth, and then follow to return to home. His use of this title suggests his appreciation for the extent to which his work was enmeshed in the environment of the place where he wrote, and many of the essays in the volume reflect his understanding of this connection. It might even be said that Heidegger's contributions to philosophy could not have been made in any other setting, or at least would have been quite different from the work that we now know. In a way, Heidegger's hut *was* the philosopher. His son Hermann seems to have understood this when in a visit to the hut filmed as a part of a documentary for television, he walks into his father's study and says movingly, "He is still alive here as far as I am concerned."[25]

Can homes that, through the investment of technology, become alive, achieve this kind of intense and intimate connection with their occupants? Can any amount of design skill and technological acumen come to replace the feeling of genuine love that we can develop for a home that we have put together, piece by piece, over long years of thought, labor, experiment, and experience? Or is it more likely that no matter how cleverly crafted the simulacrum of the responsive home as a mother's womb, the experience will feel wrong, uncanny, and slightly out-of-kilter, just as the beckoning fronds of Beesley's *Hylozoic Soil* reminded me of empathy, but left me feeling ever so slightly threatened and off-balance? And even if the answers to such questions end up being positive, is this a future that we really want?

An equally salient question concerns what it is that we have to gain through the widespread adoption of emerging technologies for connected, responsive home spaces. With our live-in proto-butlers to watch over our every need, potentially freeing us from so many of the ho-hum banalities of everyday existence, how will we use our newfound freedoms? Will they allow us to soar to greater heights? Or rather will the breaking of the chains that bind us with human connections to our homespaces make love impossible?

CHAPTER 3

PLACES OF LUST

W E CAN GROW TO LOVE a building or a place in much the same way that we grow to love a person. Enduring love develops over time and with repeated positive experiences. The history of interactions with another person nurtures feelings of trust, openness, and affection. In much the same way, our history of visits to a place, the time we spend there, and the experiences that we have, can give rise to strong feelings of attachment. And much as with love between humans, what we bring to a relationship from our own past can be just as important as what we experience in new encounters with a place. Our first experience of the Eiffel Tower or the Empire State Building is only partly colored by the raw appearance of these buildings and more by the complex of associations that we bring with us to the experience, and how that history connects with our present experiences. More volumes have been written about human love than about anything else in the world, but its mysteries still consume us.

But in our relationships with others, we're not always looking for love. Sometimes we have no interest in long-term endearments or cozy feelings of attachment. We're looking for a buzz of excitement, a momentary thrill, a jolt of human contact. We are driven by lust, pure and simple. So what is the psychogeographic equivalent of lust?

I found myself discussing such heady matters over tea with Brendan Walker in the London studio that he shares with his photographer wife and several wiry whippet dogs. Walker began his professional life as an aeronautical engineer, but early on in his career, he tired of designing military aircraft and sought to find his thrills in other domains. Now, as the self-described "thrill engineer," he spends much of his time both trying to understand where excitement arises in the built environment, and learning how to maximize thrill for those who crave a quick knee-trembler with an exciting place.

In his early work, Walker was inspired by an unusual source for ideas about how to build thrilling places: he looked at anecdotal accounts of the thrills experienced by criminals during the perpetration of illegal acts. In his book *The Seductions of Crime*, UCLA criminologist Jack Katz analyzed the motivations of various classes of criminals ranging from habitual petty shoplifters to cold-blooded killers. Many crimes, especially the more serious ones like murder, are committed in the heat of anger. But some crimes, particularly property crimes like vandalism or shoplifting, are committed purely for the pleasurable sensations that arise from the act. Indeed, some shoplifters, especially women, reported feeling orgasmic highs shortly after a successful heist. The thrill, they say, emerges from the transformation of an everyday experience (shopping) into something much more sublime and symbolic.[1]

Taking Katz's observations of the joys of crime one step further, Walker began a project called *Chromo11*, in which he invited people to visit a website where they could describe their most thrilling experiences.[2] The collection of interviews on the site illustrates the wide range of experiences that participants can find thrilling: erotically charged episodes of exhibitionism, spanking, and group sex encounters were somewhat predictable. Others involved an element of danger (reckless driving, high-risk sports), while a few were a little more off-the-wall (one respondent reported that their biggest thrill came from cracking an egg on their mother's head during an argument). In his extraordinary book *The Taxonomy of Thrill*, Walker tries to boil down the essential elements of the thrilling experience using methods similar to those employed by the criminologist Katz.[3] He deconstructs the thrilling experiences related to him by his informants into a number of separate phases, beginning with anticipation and ending with afterglow. Walker eventually reduces the thrilling experience to a set of clever equations (and what else would we expect from a thrill-seeking engineer?) that

suggest that the essential elements of thrill are not only strong physiological arousal and feelings with high positive valence, but also the rate
at which such feelings change over time. A thrilling experience is one
that jerks us quickly from the equilibrium state of our everyday experiences into something novel, disorienting, and euphoric.

Naturally enough, Walker has turned to the amusement park as a
source of inspiration, and to roller-coaster rides in particular. In his
thrill laboratory, he hooks up participants with physiological recording
devices that can measure heart rate and skin conductance so that he
can enjoy a small window into the inner workings of the body during a
roller-coaster ride. Although surges in heart rate and the clamminess of
palms can certainly indicate when a person is becoming aroused, such
measures do not differentiate between euphoric highs and plummeting
feelings of anxiety and doom. To differentiate between these types of
states, Walker employs small cameras that record the flickering record
of facial expressions as riders are thrown through the air in modern
multiple G-force amusement park rides.

Walker has been able to show that his equations, constructed on the
basis of phenomenological reports of feelings evoked by experiences
such as running naked through a suburban neighborhood at midnight
or being whipped with a riding crop during a naughty backstage frolic
at a theatrical performance, can be used to quantify the thrilling properties of an amusement park ride. His work has helped to move the
entertainment industry toward a standard measure—the "thrill factor"
that can predict how, on average, participants will respond as they are
tossed and turned through space at a theme park.

Why should we care about thrill? After all, the way that the rapid and
unexpected movements of powerful pulleys affect our senses seems to
be an extremely specialized instance of the psychological effect of place.
It has more to do with the unexpected application of forces than with
the slowly dawning feelings produced by immersion in different kinds
of spaces. The connection comes from a consideration of the manner in
which we experience the everyday spaces of our lives. Day after day, we
follow the same routes, arrive at the same destinations, and use the places

of our lives for a small number of preprogrammed routines of life. We retreat to home for rest and privacy, we move to our workplace to earn an income, and we buy groceries from the same supermarket, day after day. If this was *all* there was of life, we would barely be able to stand the stultifying dullness of things. Our conscious perceiving minds, experiencing a state of sensory deprivation, would eventually shut down completely, numbed by the monotonous routines of life into a sleep-walk. Think of the stories that you tell to your friends and family at the end of a long day. They are more likely to involve exceptions, novelty, and ruptures from the routines of life. Nobody wants to hear the story of how you walked into a coffee shop and purchased a latté. They want to hear the story of the customer in front of you who fell into a shouting rage because the soymilk had run out and who then pushed her way out of the shop, swearing and knocking over a few chairs along the way. Such exceptions, far from forming the punctuation of life, constitute its nouns and verbs. Those rare moments when the traveler encounters the unexpected, when the journey breaks down and rules are broken, are the moments when we wake up and pay attention. Such experiences, though they might not necessarily conform to Walker's "thrill effect" equations, are the ones that make us aware of our surroundings and their impact on us. What a roller coaster delivers over the course of a few seconds of gut-wrenching, adrenaline-gushing experience is really a metaphor for what we all cherish—the unexpected.

Theme parks, where most of the roller coasters that interest Walker reside, offer some lessons in place and lust. In most built settings, the main goal of the designer is to find a way to fulfill the building's main purpose while at the same time satisfying basic human needs. In a theme park, the success of the business depends *entirely* on the fulfillment of human needs for entertainment and pleasure. Although many have ascribed darker purposes to theme parks, such as the inculcation of conservative, patriarchal values, they function mainly as playgrounds for the senses and laboratories for human feeling. Many of them include the kinds of jolting rides that Walker has helped to develop, but historically they have had a wider purpose.

In the United States, the earliest theme parks were built on Coney Island after the famed city planner Robert Moses instituted zoning rules that designated the island as primarily a place of recreation and entertainment: a kind of escape valve for residents of crowded Manhattan. Rem Koolhaas, in *Delirious New York*, his brilliant exposition on the history of New York City's planning and design, described Coney Island as a proto-Manhattan, where experimental structures, sometimes made mostly of cardboard, were cobbled together with the technology of the day to provide visitors with thrilling rides, bizarre experiences, and even opportunities for perverse voyeurism. Lilliputia, a part of a theme park called Dreamland, consisted of a cardboard replica of the German town of Nuremberg. Three hundred little people, recruited from far and wide through advertisements directed at circus performers, populated the town. The residents were encouraged to set up their own infrastructure, political system, fire department, and commerce. Even more bizarre, Lilliputians were encouraged to engage in unorthodox sexual practices such as widespread promiscuity, homosexuality, and nymphomania. These encouraged practices were described as a form of "social experiment" but were really only thinly veiled efforts to titillate visitors to Lilliputia and so to raise profits.[4]

In a somewhat similar vein, the thrill ride called the Barrel of Love used technology to help visitors overcome their inhibitions and conservative values by forcing them into a moving tube that caused them to lose their footing and to tumble into and upon one another. This was a form of forced intimacy in which strangers of both sexes found themselves indecorously tangled together in sometimes very arousing postures. In this case, the use of relatively simple kinds of technology, in combination with basic elements of built form, encouraged people to behave in ways that cut against the grain of the conventional values of the day, compelling them to act out their secret desires and fantasies. in the present day, such uses of technology for social engineering have become part of an extremely sophisticated science that is being applied in almost every walk of life to make us feel, act, and perhaps most of all, spend our money in situations where better judgment might normally be expected to apply the brakes.

The state-of-the-art for theme parks that incorporate a strong techno-logical element is the Live Park in South Korea, still under develop-ment, with plans for new versions of Live Park in China, Singapore, and a site as yet to be selected in the United States. The Live Park is designed to envelop visitors in a completely immersive virtual reality experi-ence from the moment that they enter the gates. Visitors are fitted with an RFID tag—a small and inexpensive device that allows the visitors' movements and location to be tracked throughout the space—and they are invited to create an avatar for themselves in which they can custom-ize their own appearance. From this point, it is actually the avatar that engages in the themed activities of the park. Visitors to a giant immer-sive theater with building-sized screens showing 3D images and boom-ing surroundsound can watch their avatars perform and interact on the screens, where the movements of the visitors themselves set the script (and the ending) of the virtual reality performance that they experi-ence. In this park, the boundaries between the real and the virtual blur. No longer do the shows and exhibits require cardboard, wood, or any other physical materials. The story is played out in pixels controlled by the computers that track the collective activities of the swarm of visitors who are present in the site at any given time. Theme parks that com-pletely transcend the laws of physics in this manner open up a huge range of new possibilities for generating fantasy, pleasure, and thrill in their visitors, and they give guests an unprecedented degree of control over their own experience in the space. In a theme park like Live Park, the visitors themselves become the show.[5]

In *Delirious New York*, Rem Koolhaas argued that the theme parks of Coney Island, because they became so crowded and popular, meta-morphosed from a place of holiday retreat to an experiment in urban high-density living and the "technology of the fantastic." This ultimately led to many of the design principles that were incorporated into the major planning for Manhattan, where, similarly, designing an urban sys-tem that could function for a densely populated city required fantastic inventions like skyscrapers and elevators. A similar thing can be said of the world's best-known theme parks, and those that have set the modern standard for spaces of imagination and entertainment—the Disney Empire. Both of the U.S. Disney parks—Disneyland in California and Florida's Magic Kingdom, as well as Disneylands in Paris and Tokyo, share a hallmark feature that has been the subject of much discussion and debate. For better or for worse, the Disney enterprise can be considered as a successful laboratory or clinic that has focused on questions about what makes us like a place. Each of these parks greets visitors with a view of Main Street, supposedly a version of a small town U.S. Main Street at the beginning of the twentieth century. Yet, compared with what such real streets would have been like—jumbled, disorganized, unpaved, dusty, and likely littered with horse manure—the Disney Main Streets are a complete fiction. Nevertheless, there is no denying that for most visitors to the sites, these streets elicit undeniable feelings of happiness and pleasure. They serve not only as the entrance lobbies for the parks, but also as an integrated hub to which visitors return again and again during their stay, and not insignificantly, where they spend most of their money during the visit. Perhaps for U.S. visitors the appearance of these streets conjures a vision of a simpler and happier time in the United States than in more recent years. But in addition to the historical associations that might be set into play by the sight of Main Street, there is an appealing order, scale, and structure to the street.[6]

Just as Koolhaas argues that Coney Island technology of the fantastic had an influence on serious urban planning in Manhattan, the design of Disney's Main Street can also be said to have had an influence on town planning in the United States. This influence was made explicit in the Disney-designed town of Celebration in Florida, which employs similar principles to those perfected on the Main Street of the nearby theme park DisneyWorld. But unlike a theme park, which must always represent an escape from real life, Celebration was intended to function as an actual town. Although very small (current population is around seven thousand), the town was designed from the ground up to convey to residents the same kinds of pleasant feelings of old-time hominess as the theme park main streets. Residential avenues were built with wide sidewalks to encourage walking, small set-backs to encourage communion between the street and the houses that lined them, and cars packed away in garages that were only accessible from the rear of the residences. The streets are tidy, scrubbed clean, and polished to a high gloss. When Celebration was first built, homes there were in such high demand that the town held a lottery for the right to even speak to a sales agent about making a purchase. Despite the fact that some visitors report that the town conveys a Stepford-like perfection that negates feelings of authenticity, the town of Celebration can be considered a success on many levels.[7]

A Night at the Museum

Museums, most of them dependent at least in part on the public purse, are designed to educate, help shape cultural identity, and to give voice to the narratives that define our collective lives. Public museums are a relatively recent invention, having sprung from their predecessors— collections or "cabinets" of artifacts held mostly for the pleasure of their wealthy owners. Even some of the more modern ostensibly public museums, such as the famous British Museum, were in earlier times closely guarded bastions available only to the well-heeled and after making formal application for the right to enter. Now, though most of us may appreciate the value of museums as repositories of the "things" of history, fewer of us actually visit these institutions. In the most recent large-scale survey of attendance at art museums conducted in 2012 by the National Endowment for the Arts in the United States, it was revealed that only about 20 percent of Americans had set foot in an art museum or gallery in the previous year, whereas more than 70 percent of us had satisfied our yen for culture using some form of digital media. Though this survey applied only to art museums, the sunniest reports of attendance at other types of museums suggest that attendance is either flat or has been declining slightly since 2009.[8] In response to this crisis in attendance, and especially in the face of an onslaught of online opportunities that allow certain kinds of museum experiences to be enjoyed from the comfort of the living room armchair, museum curators are working hard to find ways to bring visitors through their doors. Some of these efforts are reminiscent of those undertaken by theme park designers. How do we make museums thrilling?

An important corollary to this question might be to ask when and how museums *stopped* being thrilling. As a young schoolboy growing up in Toronto, I looked forward to few school experiences more than the annual excursion to the Royal Ontario Museum. When the yellow bus stopped in front of the massive portals of the venerable old building, children peeled out of their seats in a rush and clamored to be the first through the gates and on the way to an intimate experience with ancient Egypt. There's no denying that the main attraction was not the chance to explore the fascinating and well-preserved hieroglyphs on the museum's vast collection of Egyptian amphorae, but rather the chance to peer inside a sarcophagus at the mummified remains of a human body. There was an urgency and immediacy to the experience. Even to our young, unformed minds, there was little question that we were in the presence of things that were real—that the chance to look through glass cabinets at carvings, jewelry, and pottery that was unimaginably old brought us into contact with times, thoughts, and places that were unreachable in any other way. Many of us relished the opportunity, when security guards were looking the other way, to poke our hands through a guardrail or across a barrier to make actual contact with antiquity using our hands. I'd like to think that these petty breaches of museum rules meant something more than that we were seeking a frisson of thrill like Brendan Walker's *Chromo11* witnesses. Rather, we were seeking to break through centuries of time using our fingertips to make our connection with ancient great civilizations, quite literally, palpable.

Contrast this with one of my own recent experiences as a parent of children who are about the same age now as I was during my heady adventures in the museums of my youth. On one excursion, I took my children to a local science museum that possessed as one of its prize possessions an authentic chunk of rock collected from the surface of the Moon. As we approached the exhibit, I felt a tingle of excitement not so much in anticipation of my own experiences, but of theirs. I imagined what it might feel like as a child to stand facing a genuine artifact that had been collected from outer space by astronauts. The reactions of the children disappointed me. They peered through the glass at the grey lump of rock as if they somehow expected more. The mere authenticity of the specimen didn't seem to mean much to them. More recently, I've questioned my children about their favorite museum experiences, and they've described enjoying plastic reconstructions of animal skeletons and augmented reality screens meant to show how dinosaurs, whose fossilized bones stood right before them, might have looked when they were alive. Trying to avoid leading the witness, I asked them whether the authenticity of an artifact was important to them. When looking at a pile of bones, for example, did it make a difference to them to know that they were actual fossilized bones from an animal that had lived thousands of years ago? My questions were mostly met with confusion and shoulder shrugs. Over the course of a short period, something important had changed, and perhaps not just for our children. We seem to have transformed an attraction for authenticity with one for fidelity. We're more interested in whether things *look* real than whether they *are* real. This important change in perspective has wide-ranging implications not just for our appreciation of the bones of a wooly mammoth, but more generally, for how we understand and respond to places and events.

Savvy museum curators understand that there is no prospect for turning back the clock to a time before it was possible to experience thrilling 3D reconstructions of romping velociraptors or full-motion virtual reality rides through primordial rainforests. If museums are to thrill us, then it seems inevitable that one part of their mission will be to embrace the same kinds of methods of generating thrilling experiences that Brendan Walker has pioneered for roller coasters and that others have developed for theme parks. On a recent visit to a showing of the acclaimed museum exhibit "David Bowie Is," I experienced the power of multimedia immersion for enhancing the presentation of a collection of artifacts. To have the full experience of the exhibition, visitors were more or less required to don a location-aware headset that presented snippets of interviews, music, and background sounds to accompany the visual displays. The presentation worked at the individual level by presenting me with aural entertainment that was seamlessly matched with the video or artifact I stood before, and for this type of exhibit, there can be little doubt that the power of the experience was enhanced by the technologies that were used to sculpt the presentation to my own movements. At the same time, I experienced the strangeness of undergoing a group experience in a packed venue in a state that varied from detached solitude as I peered at a youthful schoolboy portrait of David Bowie, barely aware of the other visitors to the space, to shared delight as I gazed up with hundreds of others at a large overhead display that showed rare concert footage. The artful blending of real artifacts with digital copies, personalized by the signature of my own movements through the exhibition, left me with new knowledge of the discursive facts of Bowie's life but also with a set of feelings ranging from excitement and energy to awe. What is perhaps most important about all of this is that the changing sensibilities of the audiences for such exhibitions, along with emerging technologies that are being used to track our movements and to individually tailor our experiences based on such preferences, have begun to transform the museum experience entirely. These two kinds of developments—the changes in what we demand from an exhibition and the possibilities engendered by new tools—have entered into a positive feedback loop. As we become inured to the simple power of standing before a painting by Monet or an exquisitely crafted piece of gold jewelry from ancient Rome, we demand increasingly adrenaline-soaked emotional stylings to go along with our glimpses into the narratives of civilization, both ancient and modern. Understanding how these feedback loops can be leveraged to produce more powerful and effective museum experiences will continue to constitute an important element of museum development and theory.

In the most ambitious attempt to measure the psychological state of a museum-goer, the eMotion project led by Dr. Martin Tröndle of the University of Applied Sciences in Switzerland used the same kinds of cutting-edge tools employed by Brendan Walker's Thrill Laboratory to measure the movements, gaze, and physiological arousal of visitors to an art museum.[9] Visitors to a custom-designed exhibition were invited to wear a specialized glove that tracked their location as they walked through the gallery. Proximity sensors in each of the exhibit's rooms recorded their paths, their walking speed, and the lengths of their pauses before particular objects. The gloves also monitored some aspects of the emotional state of visitors by means of records of skin conductance and heart rate. The experimenters also collected demographic data from visitors, and conducted interviews with participants so that they could assess the impact of variables such as their interest and experience with fine art on their responses to the exhibition. The results of the experiment were presented as a series of fascinating visualizations in which movement paths through the space were overlaid with records of physiological responses.

The initial study showed the viability of using such a method to measure psychological responses to museum art works, but it also arrived at some interesting and tangible conclusions about what happens to us as we walk through a gallery. For one thing, there was a strong correlation between the physiological measurements and visitors' aesthetic judgments of what they saw, suggesting that measurements of the body have predictive value for understanding aesthetic responses. Perhaps more importantly for museum curators, the experimenters also discovered some compelling differences between the experiences of solitary visitors and those who walked through the gallery with partners or in groups. The solitary wanderers, in general, experienced deeper and more frequent moments of presence and engagement with the works of art. Though it might not be surprising that visitors who were free from the distraction of companionship and conversation were more engaged with the exhibition, the quantitation of this effect with a series of carefully measured variables can be used to guide curators who want to maximize visitor engagement. In my experience of the Bowie exhibit, I was able to flow freely between a heavily internalized and intimate relationship with a single object while buffered from the crowd with my earphones, and yet a moment later I was able to join in the excitement of a simulation of a rock concert, which included the synchronized gaze and movement of a crowd of onlookers (and, importantly, my own *awareness* that I was a part of a larger group). Developing the tools that can pinpoint the rising and falling curves of such experiences, as demonstrated by the eMotion project, can only help to refine the seductive thrill factors at play in cultural institutions such as galleries and museums.

Gambling on Success

My office neighbor at the University of Waterloo is Dr. Mike Dixon. Mike is a tall, gentle, soft-spoken man with a long pedigree in research that includes work with patients who suffer visual problems after brain damage, and the remarkable phenomenon of synesthesia, in which one sees peculiar combinations of sensory properties, such as numbers that appear as colorful objects. Dixon has made groundbreaking discoveries in both of these areas; however, recently he has turned his attention to problem gambling. Visitors to casinos, or those who play at video terminals in bars, are looking for the cheap thrill that comes from the sounds and lights of a machine offering the unlikely possibility of a cash payoff. These machines are remarkably effective at removing cash from the pockets of their users to the extent that some people have become so addicted to the thrill that they have lost their possessions, their marriages, and sometimes even their lives as they spiral into suicidal despair at their inability to control their impulses. Indeed, the incidence of suicide in problem gamblers outstrips by a considerable margin the rates of suicide for all other forms of addiction. Dixon has studied several aspects of such problem gambling using generally the kinds of tools that I've been describing that allow one to monitor the activity of brain and body. Participants in his experiments are wired for measurement of skin conductance, gaze, and heart rate and they are presented with different kinds of gambling scenarios using genuine electronic slot machines. A visit to his laboratory immerses one in a crazy environment of sights and sounds reminiscent of a real casino—his experiments are the most popular in the department among our volunteer pool of psychology students. Much of Dixon's work has focused on the well-used casino tactic of the "loss disguised as a win." The electronic machines are programmed in such a way as to present to the gambler the illusion that they have won a payoff even though over a slightly longer term, they are actually losing money. Dixon has shown that these near-wins cause heart rates and skin conductance values to spike, bathing the gambler's brain in an irresistible soup of reinforcing chemicals, and egging them on to spend even more money.[10]

The sum total of my own experience with casinos came during a visit to Las Vegas several years ago with my brother. I sat before a slot machine, feeding it dollar bills and trying to make sense of the flashing displays and tinkling tunes indicating where I stood. After about a dozen tries, the machine told me that I had "won" a substantial amount, but it translated my winnings into a display showing how many gambling chips I had at my disposal. My immediate reaction was to settle back into my chair, happy that I had the chips to play the machine about forty more times. My brother, walking up behind me and watching me for a second, interrupted me. "What are you doing?" he asked. "You've won!" "I know!" I replied, "it's great, isn't it? I have lots more plays now." "You do realize that if you stop now and cash out you'll have about $200?" he responded. The thought had never crossed my mind that it was possible to convert the abstract numbers on the machine into real dollars and to walk away. Without my brother's intervention, I'm sure I would have sat at the machine until my $200 had been reabsorbed into the casino's coffers (in full disclosure, I should also mention that my brother is a professional accountant!). I left the machine, converted my chips to real money, and walked out of the casino. I haven't been back since.

This Las Vegas casino's ability to lift cash from my pocket by creating an air of unreality, detaching me from the realities of dollars and cents (or perhaps sense) is a tiny example of the larger game that is at play in such places of lust, which extends far beyond the mechanisms and programming of odds tables in slot machines and reaches into every corner of the design of the rooms and buildings of places of gambling. In Dixon's work, the emphasis is on the operation of the single gambling machine and how it influences behavior, but there is a long history of research in environmental design of casinos. Those who are in the business of building better gambling halls have conducted some of this work, whereas other investigators like Dixon focus on managing problem gambling behavior. Understandably, researchers of the latter kind who attempt to gain entry to operational casinos are treated with circumspection. Casino owners are reluctant to reveal all of the secrets by which they stack the odds against their clientele. Indeed, one Canadian researcher who was awarded a substantial grant from a government agency charged with helping to understand problem gambling was denied entry to Canadian casinos to conduct her studies and had to resort to standing outside the establishments, hoping to catch and interview patrons as they left the building.

The advent of electronic gambling terminals like the ones studied by Dixon has produced an important change in casino design. Slot machines—once mostly considered to be amusement for casino "outsiders": women, the poverty-stricken, and the casino newbies—were placed around the edges of the casino, well outside the zones of gaming tables like roulette and blackjack where the serious money was to be won or, more likely, lost. These sophisticated machines have now become the centerpiece of the casino and its chief earning technology.[11] But the design work of the casino begins by considering stages in the user experience that start long before the gambler actually takes a seat in a chair in front of a display. Decades of research in architectural design for casinos, some of it conducted by traditional designers and architects, but much carried out by veterans of the casino business and based on years of careful observation and experience, have prescribed several important principles that can lead a player to a machine.

Human beings have a deep affinity for curves. We are attracted to visual displays that contain gently undulating curves and we are repelled (and perhaps even a little frightened) of displays showing sharp edges. Such preferences, written into our DNA and predating our earliest experiences, also extend to the kinds of feelings that we experience during our own movements from one place to another. We much prefer taking a sweeping, curved route into a building or a room rather than a straight-line approach, especially if the straight approach requires us to make a hard turn from one direction to another. Although it isn't completely clear where such preferences come from, it's remarkable that we humans seem not to be the only animals that are affected emotionally by the shapes of our walking paths. Temple Grandin is a noted author and animal behaviorist who also happens to have autism. She has said that her own mental states have given her privileged access to the mental states of certain other animals, including domesticated animals used in agriculture. Grandin has argued both in her popular accounts and in her research that animals being led to slaughter in abattoirs are much less stressed when they follow curved paths rather than straight routes.[12] Her findings have led to widespread changes in the design of U.S. slaughterhouses meant to address certain animal welfare concerns about the emotional states of agricultural animals. In this case, Grandin argues that the tactic is effective because it helps to shield the animals from the view of what is ahead for them. The comparison to the gambler wandering into a casino may be apt.

The great guru of casino design is a man named Bill Friedman. Friedman, a reformed problem gambler himself, spent decades conducting careful observational studies of effective casino design culminating in a bible of sorts with the imposing title *Designing Casinos to Dominate the Competition.*[13] In his book, Friedman describes the power of the curved entryway, but also prescribes other important physical elements that he predicts will increase the cash yield of a casino. For one thing, he urges that casinos take advantage of a property called "mystery," long known by environmental psychologists to increase the appeal of a scene or place. Formally, mystery is defined as the likelihood that further investigation of a scene will yield new information. The classic example of mystery is the appearance of a winding forest trail, which leads the viewer further into the scene by promising that new vistas lie just around the next corner. Although far from the bucolic pleasure of a

country walk, Friedman argues that the same kind of physical arrangement of spaces in a casino—a set of partially occluded scenes that invite
the viewer inward—can exert the same magnetic pull on casino patrons,
which in this case will increase the likelihood that they will soon be sitting in front of a screen thrusting money into a slot. Indeed, many of
Friedman's recommendations, though he arrived at them through personal experience and careful observation of casino-goers, resonate with
what is known of general human preferences for particular kinds of
environments that are believed to have ancient and innate origins. Our
preferences for places that are high in both prospect and in refuge likely
had their origins in the advantages of selecting habitats that provided
protection from predators and invaders, while affording the possibility
of surveying our surrounding environment. Invoking much the same
principle, Friedman says that gamblers at a slot machine will be more
likely to take a position in a smaller alcove that offers some visual shelter from the larger space of the casino, but that does not completely cut
them off from their surroundings. Evidence of the use of this principle
can be found in abundance in casinos in Las Vegas, as well as in other
places. Banks of gambling machines are much more likely to be found
in small clusters that surround a small region of space than in the center
of a large, cavernous space.

In much of Friedman's work, one of his main recommendations is
that casinos be designed in such a way as to maximize the amount of
time that a gambler spends focusing attention on the machines themselves rather than the larger environment. In his view, attention paid
to the walls, floors, or ceilings of a casino represents wasted potential
profit. However, recently a new kind of design philosophy has begun
to take hold in some casinos. These so-called playground casinos are
designed explicitly to make us feel good by presenting us with appealing sights and sounds, often in the guise of large-scale simulations of
notable world landmarks. One can view a Venetian canal, sit in a French
sidewalk café, or gaze upon a large expanse of forested greenery, all
while on the way to the slots or sometimes even while play is under way.
The design philosophy of the playground casino is that environmental

interventions that increase positive emotions will encourage us to remain within the casino space for longer and to return to the casino more frequently. In addition, the large spaces, symmetry, muted color schemes, and presence of natural elements within the casino will produce the same kind of restoration from the taxing cognitive demands of game play that a city park might produce for a stressed-out urbanite. Experimental evidence from studies undertaken in simulations of casinos support these ideas. Design elements that are common in playground casinos do, in fact, produce feelings of pleasure and restoration, and participants immersed in such simulations report that they would be more likely to spend longer times in such settings. In general, the most potent combinations of design features in a casino are large-scale elements that support pleasure and restoration along with so-called microdesign elements (flashing lights, packing together of machines with different kinds of appearances) that increase the availability of information in a particular location. Interestingly, there are significant gender differences in the ways that these different kinds of design elements interact to promote problem gambling. Women, for instance, are more likely to gamble longer in settings that are less crowded, perhaps because they feel less observed in such settings. Men, on the other hand, show gambling intentions that are more or less unaffected by the crowdedness of a setting.[14]

Collectively, studies of gambling behavior in simulated settings suggest that the environment in which gaming takes place exerts a strong but subtly nuanced influence on our emotional state and that such influences are very likely to translate into increased profit margins for casinos. In contrast to Friedman's blunt-instrument approach, in which gamblers are treated somewhat like starving laboratory rats manically pressing a bar for reward, the more modern playground approach takes full advantage of what is known about the environmental psychology of pleasurable feelings. Any feelings of reticence that a gambler in a casino may feel about shelling out the next mortgage payment on a few more rounds of play can be attacked on many psychological fronts.

The Lust for Goods

We walk into casinos driven by a desire for entertainment, excitement, and for the remote possibility of a life-changing payout, and the purveyors of such establishments are expert at both providing facilities that meet those demands while ensuring that they receive as large a chunk of our income and assets as possible. In some respects, these playgrounds of lust and imagination are really about nothing more than the exchange of money for an artificial injection of pleasure into our lives. There are other places, though, where our first thought may be more of necessity than whim, but whose design is nevertheless still honed carefully to maximize profit. When we walk into a shopping mall, our simple intention might be to find a good deal on a pair of shoes or a computer game, but in many ways the design effort that is devoted to encouraging us to stay longer and spend more is no less intense in a mall than it is in a casino.

Shopping itself is an ancient activity; it has existed for as long as we have needed material goods and have had something to trade for them. In ancient civilizations, marketplaces were among the most important hubs of engagement and interaction, not all of which had to do with the acquisition of goods. Indeed, in many parts of the world, the marketplace is considered to be the most important public space in a settlement or city—in a very real sense its social capital. It is for this reason that urban activists in modern Western cities who strive to bring back value to a city's public spaces often focus on the importance of the urban marketplace. But the idea of shopping for pleasure, of spending one's disposable income on things that one *wants* rather than *needs* is a much more recent invention. The idea of shopping for pleasure and delight took strong hold in the eighteenth century in tandem with structural economic changes that produced a society in which many members had more money than they needed to fulfill the simple requirements of food and shelter. It took no time at all for those with something to sell to begin to think about how to compete for consumer cash. An important component of this global battle for a bigger piece of the consumer pie has always been the competition for the emotions, affection, and our lustful desire for shiny things.

An important stage in the development of retail marketing came with the invention of the department store—a single, large building designed with omnibus offerings of everything from clothing to food and appliances. Though it was not the first department store in the world (predated by department stores such as Le Bon Marché in Paris and Marshall Fields in Chicago), Harry Gordon Selfridge's eponymous Selfridge & Co. on London's Oxford Street was the first such store with an explicit design focus emphasizing the paramount importance of feelings of pleasure in the customer. Selfridge insisted on close contact between shoppers and the wares available at the store, fastidious customer service, and physical design elements such as comfortable furnishings, extensive glass cabinetry that made the wares easy to see, and exciting exhibitions (for example, an entire airplane was once on display inside Selfridge's), all of which, reminiscent of some of the design features of modern casinos, were intended to keep the shopper inside the store for as long as possible.

Some of the same kinds of design principles were invoked in the first shopping malls—very much an American invention and owing much to the architectural practice and theory of one remarkable individual, the Austrian architect Victor Gruen. Gruen fled pre–World War II Vienna as a young man and recent trainee in architecture and made his way to New York City, where he worked for some time as a cabaret performer. He won a small design job for an acquaintance who owned a Fifth Avenue leather goods shop in which he rethought the display elements in such a way as to break the mold of the fashion of the times—monolithic and impermeable street façades more like today's bank headquarters than alluring retail establishments. His approach was wildly successful. Following on this and a few other similar retail redesigns, Gruen made his way to Los Angeles where he started his own design firm. In a remarkably short time, Gruen found an opportunity to think on grand scale when he designed the world's first enclosed shopping mall, the Southdale Mall in Edina, Minnesota.

Gruen's design concept, based on the beautiful public arcades of his home city of Vienna, was to essentially build a new "downtown" that was free of all of the creeping planning errors of an American city. The retail area—another instantiation of an idealized Main Street like Disney's—was really only the central core of the design. Gruen also planned the surroundings of this area to include offices, residential areas, and places of recreation. Unfortunately, Gruen's full vision was not realized at Southdale, or really anywhere else that a mall was built. Southdale Mall, and most others with the same DNA, was cut off from the places where people lived and worked, only accessible by automobile, and so consequently surrounded by acres of parking lots. Despite this, Gruen's design was duplicated not only in the United States, but also in many other countries worldwide, causing Malcolm Gladwell to pronounce him the most influential American architect of the twentieth century.[15]

Most shopping malls share the same set of essential features. Malls are anchored with major tenants like department stores or discount stores at the ends, and these larger establishments are connected to one another with strings of smaller specialty stores, making a "barbell" design. Shoppers are able to find food in "courts," which are commonly in the form of large, noisy barns filled with fast-food vendors. Food courts are designed to encourage a brief pit stop for refueling rather than leisurely dining that eats up valuable shopping time. Early shopping malls were designed to be small and legible enough that an average shopper could cover all of their territory in a single visit that was short enough to avoid exhausting physical and cognitive resources, but more recent malls have vast and elaborate designs that often make it difficult for neophytes to know exactly where they are. Indoor shopping malls generally look bland and impermeable from the outside so that the wonders they contain are hidden to the outside world. Once inside, the mall patron is secreted into an entirely insulated, climate-controlled, and carefully contrived environment. Mirrors and other reflective surfaces are common; they are a design element that tends to slow walking speeds as we inspect our reflections. Sinuous walkways are common, and intersections between blocks of stores are often arranged at oblique angles. Both of these features make it more difficult for us to be able to pinpoint our location mentally with respect to the larger spaces of the mall, and the soothing curves make for the same kind of pleasant anticipation that is used in casinos and abattoirs. All of these interventions, instances of what marketers sometimes called "scripted disorientation" or sometimes even the Gruen Transfer (though Gruen, himself a committed socialist, would have gagged at the term), are designed to effect a mental transformation in the shopper. They may walk in the door of the mall looking for a pair of shoes, but they are readily transformed by such tricks into browsers—ready and willing to engage in pleasant, unfocused exploration and to entertain the possibility of buying all kinds of items that were not on any list of needs.

But where does lust come into it? Recalling that the whole point of the shopping experience, from the retailer's point of view, is to commandeer the shopper's disposable income, the holy grail of marketing is the impulse purchase. However a mall or a store is arranged, if my shoes are worn out then I'm going to find a way to buy another pair. But much of the money that is spent in malls (estimates range from about 40 to 70 percent of purchases) is doled out for items that the shopper had no intention of purchasing when entering the building. It is here that the design of the shopping space can exert powerful effects, and it can do so by manipulating the emotional state of the shopper.

Academic research focusing on the impulse buy shows that people are much more likely to succumb to such purchases when they are in a positive emotional state.[16] Indeed, psychological studies of impulsivity, an important area of study related to many different kinds of behaviors, including illicit drug-taking, food addiction, problem gambling, and unsafe sexual practices, have shown that we are more likely to act on our whims when we feel good. Shoppers who are encouraged to linger in stores for longer and who have close access to the merchandise are more likely to be tempted, but those who are brought into contact with the goods and *also* feel high levels of positive affect and arousal when they do so are the most likely to reach for their wallets. Retailers employ a wide range of tactics to heighten these positive

feelings. In some designs, patrons are encouraged by product place-
ments to place themselves into fantastic narratives where they can
imagine themselves wearing expensive clothes and jewelry. In very
large malls styled like theme parks, such as Canada's famous West
Edmonton Mall, shoppers are aroused by the complex and incongru-
ous juxtaposition of both displays of merchandise, grand facsimiles of
famous locations like Bourbon Street in New Orleans, and spectacles
such as full-sized roller coasters, live penguins, and working subma-
rines. These giant malls often include themed hotels so that shoppers
can spend a weekend or longer on the premises if they so desire.

Because of the importance of impulsivity to so many different kinds
of pathological behaviors, we have a rich understanding of the cognitive and brain states involved in impulsive behavior. In one laboratory
task used in both rats and human beings, subjects are presented with
a choice between accepting a small but immediate reward or a larger
reward that is only bestowed after a delay. Not surprisingly, those with
pathologies related to impulsivity are more likely to choose the immediate reward, and their brains light up strong foci of activity in areas that
have long been associated with addictive behaviors: the amygdala, the
ventral striatum, and orbitofrontal cortex.[17] Though most shoppers are
probably not addicted to buying in the clinical sense, it's very likely that
the same network of brain structures mediates decision-making during
a trip to the mall, and the environmental manipulations used so skill-
fully by retailers will massage the responsiveness of this neural network
to make an unnecessary purchase more likely.

As if the tools of basic environmental design were not enough to
stack the odds in favor of impulse purchases, the advent of new kinds
of technologies for more careful probing of the shopper's internal states
and preferences have provided some new opportunities for marketers to peer inside the minds of consumers. One of the fastest-growing
such areas is in the use of location-based technologies embedded in our
smartphones to track the movements of shoppers. Apple Corporation
famously keeps track of each time an iPhone user enters one of its stores
and correlates information about such visits with a customer's purchase

history. But this is almost child's play compared to some of the other tools that are now coming online. Cellphone providers in the United States and Canada may sell information gathered from their customer's phones to companies that harvest these data for useful information about the habits of consumers. This could include the paths we take through the city, the places where we stop, and even what we do with our phones while paused. In some cases, consumers voluntarily contribute to this mass of data without necessarily being aware of it. For example, fitness apps collect detailed information about when we walk, run, bike, or drive. There is a burgeoning business in purchasing and using these data to learn more about our habits.

However, it's not just our patterns of movements that have become open to big data enterprises. Following work by psychologists in the 1960s and beyond, we have learned a great deal about the relationships between the expressions on our faces and our emotional states. Paul Ekman's pioneering work, for example, showed that many facial expressions are universal across cultures and that these expressions can be quantified accurately by measurement of the movements of the facial muscles; many of these muscles are unique in the human body in that their *only* function appears to be the communication of feelings by their stereotyped patterns of contractions. Some of these contractions are fleeting, the so-called microexpressions that only appear for a few milliseconds and are virtually undetectable except to highly trained observers. Recently, the measurements of such facial expressions have been computerized. Ekman's group, working with machine-learning specialists, has developed software that can be used with ordinary personal computers and webcams to detect, read, and interpret our facial expressions. This software is attracting increasing attention in the marketing world, at this point mostly for study purposes, but there is little doubt that we will see such tools deployed at points of sale. In Russia, a company called Synqera has developed such technology for use in supermarket checkout lines. Customers who self-scan their purchases are scrutinized by a webcam that reads their facial expression, checks their purchasing history, and tailors special offers on the spot to their current mood.[18]

The overall effects of such technologies, designed to tap into the individual histories of buyers and to read out their current states, is effecting a transformation of our relationship with the built environment that is symptomatic of a wider set of changes in our psychological relationship to built spaces. Until the emergence of the Internet, much of the challenge for merchants was to attract customers to their place of business and to find ways to keep them there for as long as possible. This model for moneymaking is now being eclipsed by the ability of shoppers to make purchases anywhere using online portals, and in tandem with this change we are seeing strong declines in the construction of malls and department stores throughout many parts of the world. But now, thanks to mobile technologies, and especially technologies that track our movements, merchants are effectively able to follow us around, to sit in our pockets and purses all day long, and even in some cases, sneak a peek at our inner states. So even when we choose to leave our homes to venture out into the world to shop, a universe of possibilities for engaging our lust to purchase and consume goes right along with us.

Like lust of the sexual kind, the compulsive attractions that we have to the tempting places of our environment are not always pleasant feelings. They can be used in the most benign fashion; in an amusement park to lift us from the ennui of life's daily routine, or in a museum to draw us through its doors and into contact with cultures and authentic artifacts that might otherwise have difficulty competing with the stronger wares that shout to us. Nevertheless, the undeniable lure of place-lust can also pull us over a sharp edge from reason into seeming madness. We can spend more than we know we should in massive shopping emporia and we can even descend into a ruinous life-ending binge in front of a slot machine in a casino. As long as there have been marketplaces and gambling halls, savvy merchants have profited from a basic understanding of the psychology of the consumer and the environments in which consumption takes place. Indeed, some have argued that over the past century, wholesale changes to built settings throughout much of the world have been at play in which the spaces through which we roam have been bent roughly and forcefully into a shape conducive to promote the impulse to buy of the well-heeled and excluding those with little or nothing to trade. Our environment, including most of our public space has been commoditized.

With the advent of more refined technology that can collect and store information about our habits, actions, and feelings both at the individual and the aggregate level, we have designed an environment that can follow us from place to place and invade our innermost selves. Wittingly or not, some of these environmental adaptations, built in the service of commerce, effectively jack into our brains, accessing primitive neural circuitry that evolved to help us cope with unstable environments; however, in an environment of plenty, they can make it difficult for us to deny a base impulse to consume much more than we need or to engage in risky, potentially calamitous behaviors.

Like other elements of psychogeography, the roots of our lustful attachments to places can be found in adaptive responses to the kinds of events that we have evolved over thousands of years to anticipate and to use to our advantage. What is new is the development of both the knowledge and technological tools to take advantage of these ancient predispositions with blinding speed and laser precision.

CHAPTER 4

BORING PLACES

IN 2007, WHOLE FOODS MARKET, a chain of upscale supermarkets based in the United States with operations in Canada and the United Kingdom, built one of their largest stores in New York City's Bowery District in its storied Lower East Side. The supermarket, forming the centerpiece of a larger development called Avalonbay Communities, and including an expensive set of condominium apartments, occupies an entire city block of East Houston Street, stretching from the Bowery to Christie Street. Given the long history of protest against the seemingly unstoppable forces of gentrification in New York and many other great world cities—a struggle that has been at play in one form or another for as long as capitalism has existed—it isn't surprising that the local residents of the Lower East Side did not take the development lying down. For the well-off, the abundant availability of high-quality organic and non-GMO foods was a welcome addition to the neighborhood, but for the majority of people living in this part of New York, many of whom had roots going back for many generations to New York's immigrant beginnings, the scale of the new store, selling wares that few of them could easily afford, was seen as a symbolic affront to the historical values and traditions of this part of the city.

When I conducted research at the site in 2012, my interest in the building, though perhaps connected to the tumult over gentrification, was more pedestrian—and literally so. On my first visit to the location, undertaken to plan a series of psychogeographic studies in collaboration with New York's Guggenheim Museum, I was mostly interested in how this gigantic megastructure, plopped into a neighborhood more commonly populated with tiny bars and restaurants, bodegas, pocket parks, playgrounds, and many different styles of housing might influence the psychological state of the urban pedestrian. What happens inside the mind of a city-dweller who turns out of a tiny, historic restaurant with a belly full of delicious knish, and then encounters a full city block filled with nothing but empty sidewalk beneath their feet, a long bank of frosted glass on one side, and a steady stream of honking taxicabs on the other?

To discover the answer to this question, I designed a study in which visitors to a nearby pop-up museum site, the inaugural location of the traveling BMW-Guggenheim Laboratory, were recruited to take a walk through the city with me. On the walk, carefully designed to explore a series of urban contrasts, I led small groups from site to site and in each location, I had them answer questions delivered to them by means of a smartphone application. The questions mostly asked for simple self-assessments of the participants' emotional states and their level of excitement, but I also encouraged my participants to respond to some questions designed to draw from them some verbal responses giving some qualitative opinions regarding the sites. At the same time, I had the participants in my study wear small bracelets that measured their skin conductance—a simple but reliable window into a person's level of autonomic arousal—their alertness, readiness to act, or to pay attention or to respond to threat.

For one of the sites in the study, I used a location about midway along the long, blank façade of the Whole Foods Market. For a second comparison location, I took visitors to a site a few steps away, slightly further east on East Houston Street, in front of a small but lively sea of restaurants and stores with lots of open doors and windows, a happy hubbub of eating and drinking and a pleasantly meandering mob of pedestrians.

Some of the results were predictable. When planted in front of the Whole Foods store, my participants stood awkwardly, casting around for something of interest to latch onto and to talk about. They assessed their emotional state as being on the wrong side of "happy" and their state of arousal was as close to bottoming out as I saw at any of the sites on the walk. The physiological instruments strapped to their arms showed a similar pattern. These people were bored and unhappy. When asked to describe the site using words and phrases, utterances such as *bland, monotonous, passionless* rose to the top of the charts.

In contrast to this, people standing at the other test site, less than a block away from Whole Foods and still on the same side of Houston Street, felt lively and engaged. Their own assessments of their states of arousal and affect were high and positive. Their physiological arousal levels were high. The words that sprang to their minds were things like *mixed, lively, busy, socializing,* and *eating* (and there was lots of this going on at this location!). Even though this site was so crowded with pedestrian traffic that our experimental participants found it difficult to find a place to stand quietly to reflect on our questions, there was no doubt that they found this location to their liking on many levels. In fact, even though we didn't have the equipment to measure such things effectively, we could read the telltale signs of happiness or misery on our participants' bodies as they worked to complete the study. In front of the blank façade, people were quiet, stooped, and passive. At the livelier site they were animated and chatty. In fact, we had some difficulty reining in the enthusiasm of participants for this latter site. Our experimental protocol, requiring that participants not talk to one another while recording their responses, quickly went by the wayside. Many expressed a desire to leave the tour and simply join in the fun of the place.[1]

Although prior to our experiment nobody had thought to try to peer inside the bodies and minds of pedestrians in front of different styles of street façades, the strong behavioral effects of the simple appearance and design of a city street are well known. Noted urbanist Jan Gehl has specialized in using simple, clever, and unobtrusive observations of urban behavior in public places. Gehl has observed that people walk more quickly in front of blank façades; compared to the open, active façade, people are less likely to pause or even turn their heads in such locations. They simply bear down and try to get through the unpleasant monotony of the street until they emerge on the other side, hopefully to find something more interesting.[2]

For planners concerned with making city streets more amenable and pedestrian-friendly, findings such as these have enormous implications. They suggest that by simply changing the appearance and the physical structure of the bottom three meters of a building façade, it is possible to exert dramatic impact on the manner in which a city is used. Not only are people more likely to walk around in cityscapes with open and lively façades, but the kinds of things that they do in such places actually change. They pause, look around, and absorb their surroundings while in a pleasant state of positive affect and with a lively, attentive nervous system. Because of these kinds of influences, they actually *want* to be there. And because of such effects, many cities have carefully designed building codes for new construction that dictate some of the factors that contribute to happy and lively façades: in cities such as Stockholm, Melbourne, and Amsterdam, for instance, building codes specify that new construction cannot simply be parachuted into place. Buildings have to pass muster as being a good design match for other constructions in the neighborhood, there is a hard lower limit on the number of doorways per unit of sidewalk length, and there are specifications for transparency between the building and street in the form of clear windows with two-way views. In Jan Gehl's terms, a good city street should be designed so that the average walker, moving at a rate of about 5 km per hour, sees an interesting new site about once every five seconds. This does not happen in front of Whole Foods, nor in front of any of the other large, monolithic structures such as banks, courthouses, and business towers found in cities throughout the world.

A city planner is mostly interested in helping to facilitate the factors that, broadly considered, will make an urban setting work well. This means paying attention to the basic systems of a city—things such as transportation networks, safety, some basic aesthetics, and the promotion of public health and walkability. In other words, planners want to design streets so that they facilitate adaptive, healthy, and happy citizens who are easily able to accomplish their basic goals without impediments. But from the psychogeographic perspective, we can actually go further than these simple functional goals to ask how the design of an urban place might influence the psychological state of its occupants.

For the individual urban dweller, what are the psychological implications of getting things wrong? If city streets are designed with endless closed façades like those seen in supermarkets and bank headquarters, people might feel a little less happy and they might walk faster and pause less, but what is really at stake here? The real risks of bad design may lie less in unhappy streets filled with cars full of people who have no motivation to walk and pedestrians who won't be able to enjoy a coffee at a nice café, and more in the amassing of a population of urban citizens with epidemic levels of boredom.

In mainstream psychology, the scattered history of boredom research is peopled by individuals who were especially repulsed by the feeling. William James, one of the founders of modern psychology, said of the relationship between boredom and the passage of time that "*stimulation is the indispensable requisite for pleasure in an experience.*"[3] In more recent times, serious discussion and measurement of states of boredom and stimulation can be said to begin with the work of University of Toronto psychologist the late Daniel Berlyne. Berlyne's interests were shaped by his early experiences in the military during World War II; because he was a student of foreign languages, he was assigned excruciatingly boring code-breaking tasks in the British Intelligence Corps. After the war and motivated by such experiences, he returned to academic life at Cambridge, where he made a difficult decision to break with his studies of foreign languages to turn to psychology.

In the first part of his short career (he died young; he was fifty-two), Berlyne made many contributions to the study of human and animal motivation before turning, in his later years, to experimental aesthetics. These two fields may seem at odds with each other, but what linked them for Berlyne was his belief that one of the most primal urges, equal in importance to the drive for food or sex, was the need to seek information. In short, Berlyne argued that much of our behavior is motivated by curiosity alone: the need to slake our incessant thirst for the new. It's this need that drives us both to explore new places and to look at works of art; it is also our inbuilt urge to collect information that determines, in part, what we like when we do so.[4]

To make his case for the key role of information-seeking as a prime motivator of human behavior, Berlyne turned first to a branch of applied mathematics known as information theory. This powerful set of ideas, born in the laboratories of the Bell Telephone Company in the 1940s, was really designed to help understand the transmission of signals through wires. Information theory was applied in this context to describe the principles involved in communication under conditions of uncertainty, such as might happen when a signal being sent through a wire is partially degraded so that only parts of the message remain intact. The unit of information was described as the bit, and just like computer bits, a bit of information could range in value from zero, containing no information, to one, being filled with information. Using a few clever mathematical moves, information theory can be used to quantify the amount of information that is contained in a message in terms of bits. One of the keys to the theory is that to quantify information, one must be able to estimate the probability of occurrence of individual elements in the message. Elements that don't occur very often provide more information than those that occur commonly. Adding up all of the elements in the whole message can provide a number in bits that describes in bare formalities the information content of the message. To make this concrete, consider an example. Imagine that you retrieve a message from your voice mail. The message is quite garbled, but you can make out certain words. If you heard a message like ". . . the . . . to . . . and . . . you . ." then you would learn very little that was new. The bit value of the utterance would be close to zero. On the other hand, if you heard I'm . . . way . . . dinner . . . call . . . later," you could probably do a pretty good job of disentangling at least a part of the meaning of the message. In terms of information theory, both of the utterances contain the same number of words. The difference is that the first message contains only words that appear with very high frequency in English; they carry very few bits of information. The second message, in contrast, contains words with lower frequencies (and so lower probabilities of occurrence), so there is more information available.

It might seem like a very long reach from the technicalities of phone transmission lines to an understanding of information-seeking psychology, but there is a key connection. According to Berlyne, it wasn't just signals sent along wires that could be characterized in terms of their information content, but really any kind of object that we can perceive, including visual displays like pictures, three-dimensional objects, or even streetscapes. The power of the information theoretic approach to understanding and measuring our drive for stimulation is that it gives us a generally applicable method for measuring how much information a scene contains.

So how do we use information theory to quantify the appearance of a streetscape like that found in front of the Whole Foods store in New York? Think of yourself walking along such a street. As you take the first step, you see on your right a wall of frosted glass and on your left the busy street. Take another step. There's nothing new. Step three. Nothing changes. For a span of about two hundred steps, you could have predicted what you would see next based entirely on what you have just seen. Nothing has changed. No information has been passed and your nervous system is completely unaroused and uninformed, much like the receiver of a phone message containing nothing more than words like *and* and *the*. And you don't really even need to walk along the Whole Foods façade to see this. Instead, you could stand across the street from the façade, take in the whole thing at once, and see that it consists of a single monolithic slab of built space that is virtually the same everywhere. It contains a very small number of identifiable elements repeated over and over again.

Now the reason for the dismal recordings of happiness and arousal in participants standing in front of blank façades should be a little clearer. These constructions don't work at a psychological level because we are biologically disposed to want to be in locations where there is some complexity, some interest, the passing of messages of one kind or another. And this urge runs much deeper than a simple human aesthetic preference for variety. The urge to know is written into us at a very primitive level, one that we share with any other animal whose behavior has been studied. Many of Berlyne's early studies of the influence of complexity on arousal and motivation were conducted with rats rather than with people. Like us, when rats are free to explore a laboratory environment, they will consistently choose paths that lead them to areas of higher complexity and novelty. Even in the simplest of laboratory maze scenarios, the Y-maze, in which rats are released in a straight runway that ends with a single Y-intersection, rats tested repeatedly will spontaneously alternate between left turns and right turns even when there is no particular reason to do so other than to explore the route not taken. In other similar studies, even cockroaches have been shown to exhibit such preferences.

Berlyne's observations of the preferences of rats and people for visual complexity in our environment turn out to be somewhat more nuanced than to suggest that we are perpetually driven to seek higher and higher levels of complexity, chaos, and novelty in our environments. Anyone who has shriveled under the bright lights and blaring noise of Times Square in New York, Shibuya Crossing in Tokyo, or parts of the Las Vegas Strip will already know the truth of this. What Berlyne observed in his early studies, and what has been substantiated by many subsequent experiments, is that there is a kind of sweet spot for complexity. We may suffer from boredom when we walk down those long zero-bit roadways in suburbia or in the central banking districts of major cities, but we can also be unpleasantly overloaded by too much of a good thing.

Psychologists have long been interested in boredom. Though we may not all agree on a precise definition, some of the signs are well known: an inflated sense of the inexorably slow passage of time; a kind of restlessness that may manifest as both an unpleasant and aversive inner mental state but also with overt bodily symptoms in the form of fidgeting; postural adjustment; restless gaze; and perhaps yawning. But how do we classify boredom as a psychological state? Is it an emotion? A cognitive state? Something else? And how does boredom fit with arousal? Some researchers have suggested that boredom is characterized (and perhaps even defined by) a state of low arousal. Indeed, in some experimental studies of boredom, it does seem that when people are asked to sit quietly without doing anything in particular, presumably a trigger for boredom, their levels of physiological arousal do decrease. But Berlyne, and recently some others, have suggested that boredom may sometimes be accompanied by high states of arousal and perhaps even stress.[5]

In recent research conducted by University of Waterloo cognitive neuroscientist James Danckert in collaboration with his student Colleen Merrifield, participants were brought to the laboratory, hooked up to equipment that measured their heart rates and their skin conductance, and asked to watch some videos. The videos were carefully calibrated to elicit emotional states of one kind or another. In one video, designed to elicit sadness, a heartrending scene from the movie *The Champ* was shown. Another video, designed to elicit boredom, showed two men hanging laundry on a clothesline. The men simply passed clothespins back and forth to one another and hung clothes. Not surprisingly, the participants self-reported being saddened by the clip from *The Champ* and bored (or sometimes confused) by the laundry video. What was more interesting was that the two videos elicited two different patterns of psychophysiological signatures from participants. Compared to sadness, boredom elicited rising heart rates and decreasing levels of skin conductance. Ordinarily, as in other studies using skin conductance measures, one might interpret the lowered values here as suggesting that bored participants were experiencing lowered levels of arousal. However, Merrifield and Danckert included a third important measure in their study. At certain phases of their experimental procedure, they asked participants to contribute saliva samples that were later analyzed for the presence of cortisol, an important stress hormone whose levels in the body marks activity in a brain system known as the HPA or hypothalamo–pituitary–adrenal axis. Remarkably, after a brief 3-minute exposure to a boring video, participants showed increasing levels of salivary cortisol compared with levels seen when they viewed a sad video.[6] Chronically high cortisol levels have been associated with a range of human stress-related ailments, including stroke, heart disease, and diabetes.

The discovery that even brief boring episodes can increase levels of debilitating stress fits well with other recent suggestions that there may actually be a relationship between boredom and mortality rates. In a large, long-term study conducted in the United Kingdom and begun in the 1970s, participants were asked to complete a series of questionnaires, some of which asked about their state of boredom with their lives and their work. In follow-up work completed in 2010, it was shown that those participants who had reported higher levels of boredom in the earlier assays were significantly more likely to have died before the second study.[7]

Boredom does not simply force us to undergo unpleasant states of fidgeting or increased levels of bodily stress hormones. It can also lead us to engage in risky behavior. Surveys among those who suffer from addictions, including both substance and gambling addictions, suggest that levels of boredom are generally higher in such groups and that episodes of boredom are one of the most common predictors of relapse or of risky behavior such as unsafe needle use or sexual practices.

The findings of Merrifield and Danckert suggest that even exposure to a brief, boring experience is sufficient to change the brain and body's chemistry in such a way as to generate stress. This finding alone lends some neuroscientific weight to the argument that designers of the built environment have reason to attend to factors that might contribute to boredom and that the influence of environmental complexity, as demonstrated decades ago in Berlyne's pioneering experiments with rats and people, might actually affect the organization and function of our brains. It might seem extreme to suggest that a brief encounter with a boring building might engender serious hazards to one's health, but what about the cumulative effects of immersion, day after day, in the same oppressively dull surroundings?

This question has long interested psychologists, especially following Canadian psychologist Donald Hebb's original discovery that rats who lived in enriched environments were markedly superior intellectual beings than laboratory rats living in more Spartan sur-roundings. Hebb's enriched rats could solve more complicated maze problems in shorter times than their less-fortunate labmates. In his initial discovery, Hebb compared rats that had been raised by his chil-dren as pets with rats that lived in his laboratory. More controlled studies that compared the behavior of rats living in luxury condo-style environments within the walls of the laboratory showed that these rats enjoyed a favorable edge compared to rats living in standard "shoebox" cages that were the norm in the 1950s and before. Later work carried out by Berkeley's Mark Rosenzweig showed that such enriched rats were not only superior performers, but that they also had a thicker neocortex with more richly developed synaptic connec-tions between brain cells. Indeed, this finding was the cornerstone of the modern view in neuroscience that the brain, far from being a fully formed and immutable organ by adulthood, could show dramatic physical responses to environmental changes all through the lifespan (and it's one of the reasons why so many of us place such great hope in the possibility that crossword puzzles and brain-training games like Lumosity will enrich our brains so as to stave off cognitive decline as we age).[8]

But what about people? The brain mechanisms responsible for the enrichment effects discovered by Hebb, Rosenzweig, and legions of other researchers are so fundamental that it would be an extraordinary thing if these principles did not apply to us as well as to laboratory rats. And, indeed, different kinds of experiments looking at the influ-ence of skill development on brain organization have shown that our brains possess a remarkable degree of plasticity. To give one example, musicians who have engaged in demanding practice of manual skills required for performance show measurable increases in brain activ-ity and connections in areas of their brains related to the skills they have developed. It's fortunate that there are very few parallels in human neuroscience that can be compared to the experiences of Hebb's rats living in tiny cages faced only with four enclosing steel walls. When "natural" instances of such cases are discovered, they often involve children who have been incarcerated in their homes by abusive parents for extensive periods, and have typically involved enormous emotional stressors and dietary impoverishment in addition to environmental deprivation, so don't make equivalent comparisons. Perhaps the closest human comparison to the life of a typical laboratory rat in the 1940s is the human prisoner in solitary confinement. (Today's lab rats are generally housed in environments that include some environmental enrichment like that used by Hebb, Rosenzweig and others. Indeed, the findings from these early studies are largely responsible for this humane change in the treatment of experimental animals.) Even after brief periods in "solitary," prisoners who do not evince any prior mental pathologies experience long-lasting delirium, impulsiveness, and self-destructive behavior. But such experiences go well beyond a simple restriction of environmental variety. Prisoners in solitary confinement also undergo complete restriction of social stimulation, and this is probably at least as important as the four enclosing walls of the prison cell in producing the deleterious effects of such treatment.[9]

A better benchmark for the effects of environmental deprivation on behavior and brain function may come from studies devoted to pinpointing the causes of human disorders such as attention deficit hyperactivity disorder (ADHD). Here, comprehensive studies of the home environment of children have shown that the lack of availability of enrichment in the physical environment of the home, in the form of affordances for play and stimulating wall panels and artwork, is one of the strongest predictors of the symptoms of ADHD.[10] This finding fits intriguingly well with the results of the Merrifield and Danckert study because the psychophysiological signature of boredom that they identified has also been seen in children diagnosed with ADHD. Collectively, studies of both extreme and more moderate forms of environmental deprivation provide compelling evidence that boring environments can generate stress, impulsivity, lowered levels of positive affect, and an increased likelihood of maladaptive risk-taking behavior. At this point, we simply don't know the extent to which such effects might be produced by simple daily exposure to poorly designed urban environments or by building interiors because the studies have not yet been done. However, based on well-understood principles of neuroplasticity and on what is known of the effects of deprivation and enrichment in other more extreme settings, along with studies like those conducted by Jan Gehl and by my group in several cities worldwide, there is every reason to believe that these sterile, homogeneous environments are exerting a measurable effect on our behavior, and likely our brains as well. Given this, the prudent design of city streets and buildings, taking optimal levels of factors such as visual complexity into account, goes beyond the simple idea of promoting walkability and active and vibrant downtown neighborhoods. It is a matter of public health—mental health in particular.

Even without the findings from sophisticated psychological experiments and studies of the bored brain, we all know from personal experience that boring environments are unpleasant. So why do such environments happen? Why would anyone think it a good idea to build a large building that was featureless at ground level? What motivates a developer to erect an endless stretch of suburban housing where each individual unit is identical, and in the language of information theory, low in entropy? The answers to these questions are many and complicated, and at least some of them are beyond the ambit of psychological theory. One obvious part of the equation, especially for suburban developments, is the economic one. It's much less expensive to design only three or four different models of houses, perhaps with minor variations, than it is to offer a rich collection of different types of buildings for consumers, and the resultant savings are passed along to consumers to make housing affordable—an important goal. Individualizing cookie-cutter houses often comes down to the imagination and the pocketbook of consumers, who might try to make a property their own by means of design flourishes or artful landscaping. But all too often, the budget required for this is lacking in new homeowners who are already stretched to the limit in finding the resources to purchase a property at all.

What about our larger institutional buildings? Why build a closed, ground-level façade that will bore the passersby? The economic argument is somewhat less convincing here because many of the greatest sinners in this regard are corporations that seem flush with the design cash that could easily turn the bottoms of their buildings into engaging streetscapes. One possibility is that the owners of such properties may not see much to gain. It hardly seems to be in the best interests of a major bank, for example, to attract a crowd of happy lingerers to the fronts of their buildings, rather than serious customers who will get in and then get out again. A friendly façade might also be less in keeping with the image that the business would like to portray to customers. We might not want the bank that we hope is looking after our assets to portray itself as part of a whimsical and lively street market rather than as a quiet, brooding, and impenetrable fortress.

Considerations such as these—economics, image, and a reluctance to include design features that may run counter to the functions of a building (mostly security and efficiency) provide some straightforward reasons why our urban streets might not always hit the sweet spot of complexity. But there are at least two other reasons why such designs fall short of our psychological needs.

One of these reasons has to do with a radical shift in architectural design in which entire building envelopes became signs. This trend was first identified in Robert Venturi's controversial studies of the architecture of the Las Vegas strip in which building façades have famously become advertisements for their contents, but it can be seen everywhere.[11] Think of the building façade of a corporate chain establishment such as a McDonald's restaurant. In such cases, we can easily recognize the building's brand from a distance and at high speed (important when driving a car), and there is little doubt that this rapid recognizability is a part of the objective of the design. Many a weary or culture-shocked traveler (myself included!) has experienced a palpable sense of relief at the end of a hard day of travel in unfamiliar territory at the sight of such a landmark-cum-building. Entire typologies of buildings have been designed to rely on this instant recognizability to attract traffic. Supermarkets, banks, restaurants, department stores, and many other types of retail establishments rely not just on explicit signs bearing their trademarks, but also the very shapes of their building envelopes to make it easy for travelers to recognize them. Indeed, my children and I often play a game where we try to guess the identity of a new construction based on the earliest signs of its shape and size as it rises from a new excavation. It isn't a hard game to win.

Another factor in the genericization of the built environment, as pointed out by architectural historian Sarah Goldhagen, falls on the client's side.[12] Especially in North America, we suffer from a lack of understanding and appreciation of good architecture, which is engendered by a lack of education. Although we may bemoan the losses in fine arts and design curricula, often the earliest victims of austerity cuts, we have complained much less about the fact that there has never really been a serious attempt to include material on architectural design in such curricula, even though most would probably agree that it is this kind of design that is most likely to have an impact on our day-to-day lives. We might shake our heads with dismay when a hideous new public building appears, and the wide gulf that apparently separates the sensibilities of architects from those of the common user of built spaces will receive some airplay following the appearance of a monstrosity, but the root causes of the disconnect—a failure of the average person to engage in the dialogue of city-building because of a lack of understanding whose origins lie in our basic education—is much less often addressed.

On the side of architectural schools, Goldhagen also points to an inherent bias in such programs that emphasize postmodernism—a set of ideas difficult to pin down precisely, but nevertheless generally associated with a mistrust of the role of major institutions in our lives. If buildings in the public realm are entrusted to architects who are educated to hold such biases, can we really wonder at the result?

A final reason for the genericization of urban settings relates to our increasing reliance on digital technologies and information to mediate our relationship with the built environment. Such technologies produce connectedness without proximity and can emphasize the virtual at the expense of the real. To understand what this might have to do with a boring downtown streetscape, one only has to spend a few minutes standing on any urban street corner. The focus of human attention has shifted palpably downward into the upturned faces of our phones and away from our physical surroundings. Indeed, this problem has become so acute that at New York's busiest and most dangerous intersections, transportation planner Janet Sadik-Khan has ordered large, attention-catching graphics to be painted onto the sidewalk to remind distracted pedestrians to look up from their devices to avoid impact with oncoming vehicles.[13] Though this new behavior of ours may seem to be nothing more than a simple change in posture and gaze, it is one that has changed the manner in which we use city streets, and has motivated a change in design. But it is also symptomatic of more profound change: we may no longer care nearly as much about what our surroundings look like because we are not paying attention to them as we used to. In a very real sense, we are no longer *there* as we used to be, and our physical surroundings are no longer as *real* as they used to be.

The trend toward hybridization of real and virtual spaces in urban environments also has ideological roots. Indeed, though some are touting the new trends toward wired cities and the Internet of Things as ushering in the bare beginnings of a new kind of merger between information technologies and architecture, this trend has actually been under way for some time. Just as electronic connectedness enables globalization by discounting the importance of physical space and dimension in many of our everyday dealings with life, the homogenization of architectural design parallels this trend in the arena of bricks, steel, and concrete. Indeed, in their opus *S, M, L, XL* Rem Koolhaas and Bruce Mau vaunt empty-box designs in their argument for what they call "the generic city."[14] The authors argue that any kind of architectural ornament, be it a particular kind of façade design, the idiosyncratic arrangement of streets, or specific elements of cultural iconography, is destined to be, in some sense, exclusionary. They contend that in a world in which we are being thrown together into groups that transcend old cultural borders, any kind of design that contains historic associations will inevitably alienate people who do not share the histories of a particular style of design. In an interview published in *Der Spiegel*, Koolhaas puts it like this:

> The traditional city is . . . occupied by rules and codes of behavior. But the generic city is free of established patterns and expectations. . . . Some 80 percent of the population of a city like Dubai consists of immigrants. . . . I believe that it's easier for these demographic groups to walk through Dubai, Singapore or HafenCity than through beautiful medieval city centers. For these people, (the latter) exude nothing but exclusion and rejection. In an age of mass immigration, a mass similarity of cities might just be inevitable. These cities function like airports in which the same shops are always in the same places.[15]

Koolhaas' argument that the cultural trappings inherent in traditional cities may be alienating for mobile, multicultural residents of the modern city has merit. He may well be right about the inevitability of generic, functional design in an age of globalization, instant communication, and ubiquitous connectedness. However, unless it is possible for our electronic connections with the world and with one another, brought about by the panoply of technologies that are now available for communication and simulation, to completely supplant our physical surroundings the widespread adoption of global, universal, functional designs will have psychological consequences of the kind that I've described in this chapter. It seems almost inevitable that the genericization of built spaces will reduce their variety, interest, and complexity. Human beings have evolved to operate in particular types of environments, with optimal levels of complexity that are ultimately related to our biology. We seek out such settings with our eyes, our bodies, our hands, and our feet, and in turn, the design and appearance of those settings, by affecting our bodies, tap directly into ancient circuits meant to produce feeling responses and emotions that are adaptive. Very little of this is cultural. Such settings tune us affectively to our surroundings, help us to maintain preferable states of arousal and alertness, and ultimately permit us to behave adaptively. In a generic design where our transactions with our surroundings are mediated entirely artificially using embedded intelligence and carefully designed interfaces that can simulate the environmental contingencies for which we evolved, it *might* be possible for us to get everything exactly right and to produce the perfect human environment. But given the complexities that we would need to understand and to model in order to do so, it seems more likely that we will get enough things wrong so we will be worse off than ever before. Furthermore, producing the perfect, adaptive environment out of bits and pixels would also seem to presuppose the oversight of an entirely benevolent and unbiased set of powers. Given what we know of the artful use of design psychology by those with vested interests to influence our behavior to increase their own profit, up to and including the production of damaging and self-destructive behaviors akin to those seen in people with substance addictions, this supposition of benevolence would seem to be naïve.

As much as we might like to have it be otherwise, it seems that boredom is an inevitable element of modern life. One might even argue that *some* degree of boredom is healthy. When the external world fails to engage our attention, we can turn inward and focus on inner, mental landscapes. Boredom, it has sometimes been argued, leads us toward creativity as we use our native wit and intelligence to hack boring environments to create interest. But streetscapes and buildings designed and built to generic functional requirements and ignoring the inbuilt human need for sensory variety—a tempting and economical proposition when so much of our mental stimulation comes from the virtual and the electronic—cuts against the grain of ancient evolutionary impulses for novelty and sensation and will not likely lead to comfort, happiness, or optimal functionality for future human populations.

CHAPTER 5

PLACES OF ANXIETY

O N November 28, 1942, 492 people died at the Cocoanut Grove night club in Boston after a fire broke out when a busboy lit a single match to retrieve a light bulb that he had dropped on the floor. In the ensuing panic, patrons swarmed toward the single exit from the building, a revolving door at the main entrance that soon became blocked by the crush of bodies attempting to flee the site. In terms of the mortality rate, the Cocoanut Grove fire was the worst nightclub fire in history, but it stands far from alone as an instance of the disastrous influence of human emotion interacting with the physical layout of an environment and causing death.[1]

Every year at the Hajj, a massive pilgrimage to Mecca involving more than five million participants, several hundred people are crushed underfoot when the crowds, reacting in part to the high emotion of the event, challenge the limitations of the physical setting. Sporting events, street festivals, parades, and demonstrations sometimes similarly turn awry, causing human tragedy. Such instances of death by crowding have generated a small industry in preventive measures ranging from building codes that impose upper limits on occupancy and lower limits on exits to sophisticated modeling software in which movements of crowds of individuals can be predicted in proposed designs before they are constructed. Nevertheless, despite such measures, the simple and important fact remains that the level of anxiety experienced by visitors to any built setting will influence their behavior both at the individual level and at the aggregate level of the crowd.

When we feel threatened, our natural response is to seek the quickest way to remove ourselves from the site of threat; this most often means that we will try to find the most direct route to the nearest exit that we know about. When that route is not available to us, we may become injured and unable to move at all. Although I've described some of the most dramatic and terrifying consequences of anxiety in a built space, the same kinds of effects, but with more moderate outcomes, are operating on us constantly as we move from place to place through a building or a cityscape. Such reactions can be seen as the counterpoint to the kinds of responses that may attract us to particular locations and make us feel happy and stimulated when we are there.

Anxiety can be normal and adaptive. Psychologically, it arises when we anticipate some kind of unpleasant future event, which could be anything from a physical injury to the unwanted exposure of our inner state to the view of strangers. Just as it is a healthy response for us to withdraw a hand quickly from a hot stove to avoid a painful burn, it makes sense that we would have inbuilt mechanisms designed to manage more complicated responses where we must assess an environmental threat and respond to it efficiently. In built settings, difficulties arise when effective responses to perceived threats, for one reason or another, are not available to us. In such cases, we are forced to dwell in uncomfortable locations of high, perceived threat, giving rise to a cascade of neural and endocrinological responses that may produce mental pathologies or decrements to our physical health.

Urban Pathologies of the Mind

For reasons that are not yet well understood, psychiatric disturbances related to anxiety occur with greater frequency in urban environments. Diagnosed rates of anxiety disorders, clinical depression, and schizophrenia are all higher among those who live in cities than among those who live in rural areas. Differences in socioeconomic status, the possibility of exposure to toxins or pathogens, and many other types of environmental threats peculiar to urban settings have been considered as possible explanations for the higher rates of psychiatric disorders in cities, but no such accounts have been found to be particularly convincing. Some studies have suggested that sociological factors such as neighborhood cohesion (or more accurately, the lack thereof) may be responsible for rising rates of these afflictions in cities, and here the evidence has been more compelling. For example, a detailed investigation of neighborhood cohesion, including measures of neighborhood mobility, incidence of single-parent households, and family size suggested that more cohesive neighborhoods were less prone to increased rates of anxiety and depression.[2] Such findings are important because they suggest that the factors that control rates of these diseases may be amenable to modification by means of urban design measures. A particularly interesting finding suggests that the availability of natural spaces in an urban environment may mitigate the risk of psychiatric disorders in urban populations, again suggesting that the tools to help lessen the mental toll of city life may be in the hands of architects and urban planners.[3] However, until we understand more about how urban life triggers psychiatric ailments, design interventions that might be instituted to mitigate such risks are likely to be hit-and-miss.

In a groundbreaking study on the effects of urbanization on brain activity, a joint German–Canadian team of neuroscientists led by Dr. Andreas Meyer-Lindberg of the University of Heidelberg made a startling discovery.[4] In their brain-imaging experiment, they asked participants difficult mathematical questions while their brain activity was being visualized and—this was the key variable—while they were also subjected to negative social stress by receiving feedback from the experimenters suggesting that their performance on the mathematics task was much worse than expected. In other words, participants received false feedback about their performance, being told that they were making errors even when they were doing well at the task. Not surprisingly, participants in the study showed strong signs of stressful arousal according to measurements of their sweat gland activity. In this case, the source of anxiety is clear—participants were made to feel that they had exposed an important part of their inner selves—a deficiency in mathematical ability. What was most interesting of all, though, was that the patterns of brain activity produced by the procedure varied depending on both the residential backgrounds and the current living arrangements of the participants. Those who lived in large cities showed stronger activation in the amygdala than those living in smaller settings. Those who had been *raised* in large cities showed stronger responses in their cingulate cortex than those who had been raised in rural settings, regardless of their current living situation. Both of these brain areas, the amygdala and cingulate cortex, constitute important parts of the neural pathways by which we respond to emotional events, especially threatening ones, and learn associations between environmental contexts and their probable outcomes. The importance of these findings is that they suggest that those of us who were raised in or who live in large cities show larger brain responses to social triggers for anxiety than those of us who live in rural settings; so the findings provide an important clue to the locus of one possible effect of city living on our brains. Results from this study suggest that crowding, in which we are forced to live among large numbers of strangers, generates a set of social stresses on us as we attempt to adapt our own patterns of behavior to those around us. The brain-imaging findings further suggest that these chronic social stresses can actually change the ways that key brain areas involved in the mediation of affect and stress respond to acute social stressors.

Geotracking Psychiatry

Ultimately, unraveling the complex relationships between the urban environment and behavior would require a much more detailed knowledge of the day-to-day transactions that individuals living in cities undertake as a normal part of their lives. But now, new methodologies based on geotracking smartphones may make such studies possible. Jim van Os of Maastricht University in The Netherlands is developing methods whereby participants can be tracked as they move through their city so that researchers know exactly when they reach a potential stress-point—a noisy and crowded train platform or a busy marketplace, for example—and participants can be polled on the spot to report their feelings.[5] They can even receive some cognitive testing *in situ* so that their ability to cope with their current circumstances can be assessed in the field. Such methods, combined with the brain-imaging methods used by Meyer-Landberg, might be used to reveal the moment-by-moment unfolding of the effects of stressful urban experiences on the individual psyche.

The work that has been inspired by Meyer-Landberg's group may eventually go far to explain at an almost microscopic level of detail the manner in which urban pathologies are triggered by our unnatural living arrangements in congested cities. But what is perhaps even more tantalizing than this is the possibility that to the extent that such work could describe the detailed architecture of the environment–brain relationship, it might be possible to reverse-engineer our urban habits in such a way as to provide a therapeutic intervention for harried city dwellers. Some current smartphones are already capable of constructing geofences, geographically defined zones that trigger our phones to carry out simple commands. At present, the most common use of such geofences is to provide us with reminders; our phone might know, for example, when we are leaving our offices and prompt us to remember to stop at the market to purchase some groceries. But if it were possible to construct an individualized stress map of one's common destinations within a city, the geofencing functions of our phones might also be tweaked to warn us when our daily exposures to stressors have exceeded our maximum recommended daily allowance. The phone would function somewhat like the film badge that workers who are exposed to radioactivity use to monitor their daily dosages of gamma rays to keep them at acceptable levels. Except that in this case, the devices would be programmed to understand our habits and vulnerabilities, and to warn us when it's time to make a healthy retreat to a quiet zone or a healing green space.

In a recent discussion I had with Ed Parsons, a geospatial engineer and the self-described "good will ambassador" for Google's European arm, he left no doubt that the individualization of our personal relationship with our geographies, so that the world that our devices presents to us is closely aligned with our own preferences, is a major goal of the company.[6] Although I blanch slightly at the prospect of the widespread adoption of such individualized maps and the loss of community that might result, in this particular case it is hard to deny the appeal of a kind of emotional prosthesis that sufferers of a serious psychiatric illness might use to help keep them out of the way of environments that will worsen their condition. A more thought-provoking question concerns the extent to which the general public, most of whom have the resilience to weather the kind of urban turbulence needed to navigate through a typical day in the city, have anything to gain from a device that might encourage them to geofence themselves off from the world's sharper experiences. When does a *prosthesis* designed to ease suffering and help us to cope with damaging psychological distress become an unnecessary *crutch* that may prevent us from the full-on experience of life's more unexpected and intense peak experiences?

A skeptic might also argue that we already know when we are stressed out and that the problem is hardly one of knowing when to flee for refuge as it is finding the opportunity to escape the pressing exigencies of daily life. Yet there are reasons to doubt that we do have the emotional acuity to avoid urban stresses, even when we might be able to do so. For example, when I visited Mumbai to conduct psychogeographic measures of human–environment interactions, I was astonished by the intricate dance (something like a combination of ballet and parkour) required for a pedestrian to cross congested streets safely when vehicles rarely came to a full stop and drivers rarely ceased leaning on their horns. When I asked local residents how they coped with such daily adrenaline baths, they commonly responded with quizzical looks and shoulder shrugs, usually saying something like, "Oh, it's not a big deal. You get used to it." To put them to the test, I placed groups of pedestrians in the middle of busy traffic intersections and I measured their physiology directly. Though my psychological measurements suggested that my participants were not very happy compared with their feelings while standing in a nearby garden, their own self-reported arousal levels were unexceptional—very much in keeping with the anecdotal accounts they had given to me. On the other hand, their sweat gland responses were off the charts, suggesting that although they might feel that there was nothing particularly unusual or taxing about being buffeted by a swarming sea of noisy cars and motorcycles, their bodies were showing stress reactions that were probably not very different from my own. Human beings are remarkably resilient. We can adapt to a wide range of environmental circumstances, even those that are unpleasant in the extreme. But just because we can exhibit such psychological resiliency does not mean that, behind the scenes, the stress-responsive parts of our bodies and our central nervous systems are not doing their work, raising our blood pressure, filling our bloodstreams with cortisol, and perhaps ultimately making us more vulnerable to both physical and mental illness.

Despite the tantalizing links between chronic exposure to the unpleasant side of urban living and our reactions to social stresses in laboratory experiments, it's fortunate that only a minority of those who live in cities suffer from full-blown psychiatric disorders. This suggests that it takes more than the daily buffeting of crowds, horns, and the occasional insult hurled by a stranger to cause deep depression, generalized anxiety, or psychosis. As it has long been known that these psychiatric disorders are, to some degree, heritable, another key part of the equation is likely to be genetic in origin. But even here researchers have begun to zero in on some key gene–environment interactions. One particular gene, which provides the blueprint for one type of receptor for a neurotransmitter known as neuropeptide S shows a relationship with the genesis of stress-related psychiatric disorders. Not only is the presence of this gene related to greater proneness to pathological stress reactions such as acute anxiety, but in experiments much like the brain-imaging studies comparing city and country dwellers, individuals carrying the suspect form of the neuropeptide S receptor *also* show heightened amygdalar responses to social stress, just as was seen in the long-time urban dwellers in the earlier study.[7] This intriguing finding strengthens the link between urbanicity and psychiatric diseases by suggesting that although we all may show brain changes in response to urban stress, only those carrying the right kind of neuropeptide S receptor gene will be strongly at risk for a full-blown disease state.

If further research bears out these early findings, then the possibilities for using individualized stress maps carried by our pocket technology to mitigate our exposure to harmful urban settings could be extended further still. A stress map based on our genetic vulnerabilities could augment an empirical stress map based on responses actually recorded in our familiar environments. In an age where individualized genetic fingerprints are becoming increasingly inexpensive and easily available to consumers, such notions are far from science fiction. Perhaps the more important issue is how far such a detailed map of our psychological responses to place, based both on our past experiences and our genetic makeup, should be permitted to guide our movements through the world.

Shape Matters

The research we have looked at so far suggests that there may be features of our built surroundings that provoke anxiety and that repeated exposure to these architectural elements may cause changes in our brains that make us more reactive to stress. Some of us, perhaps those with genetic predilections to pathologically high arousal reactions to stressful events, may even develop serious psychiatric disorders as a result. However, so far we've made little headway in identifying exactly which features of urban surroundings are responsible for such effects. There are some obvious candidates, such as exposure to excessive noise, which has long been known to affect cognitive performance and emotional state in every kind of setting from the factory floor to a busy street corner. There is also a kind of sweet spot for sensory complexity, which when exceeded prompts us to take measures to avoid and seek shelter. It would also make sense that chronic exposure to the ordinary risks of urban environment such as hazardous traffic would also serve to heighten anxiety. But apart from these obvious kinds of disturbances to our city equilibrium, it's likely that the very shapes of the buildings that surround us can also contribute to urban unease.

In 2007, the city of Toronto saw completion of a controversial new addition to its venerable old Royal Ontario Museum, a stately, Italianate style building originally built in 1914 and expanded, but in much the same kind of traditional architectural style, in the 1930s. Designed by deconstructivist architect Daniel Libeskind, the more recent addition, named the "Lee-Chin Crystal" in honor of a major benefactor, consisted of a large volume of glass- and steel-enclosed space containing sharp contours and a vertigo-inducing lack of balance and orientation. Although some critics hailed the construction as a masterpiece heralding a new age of cutting-edge architectural design in the staid, Victorian landscape of Toronto, the building was decried by many as an abomination born of the city's urgent desire to establish itself as a world-class destination by employing the services of a globally acclaimed "starchitect." Casual passersby deplored the transformation of the public outdoor space surrounding the museum as a result of what one commentator described as a "bizarre alien turd" having been dropped into the landscape. Pedestrians avoided the open, windblown site lying under the Crystal, which appeared to teeter from above. Many expressed concern—not completely without warrant—about the possibility of dangerous icicles falling from the sheer angles and sharp edges of the building during the fraught Canadian winters. Some long-time devotees of the museum claimed they would never set foot in the building again. In 2009, the Crystal was voted by members of virtualtourist.com, a major online portal of information for world travelers, as the eighth ugliest building in the world, and in the same year the *Washington Post* named it the ugliest building of the decade. Many harsh critics cited the uselessness of the new space, despite the fact that it allowed the museum to put on display countless artifacts that had previously been hidden away in storage. Few of these critics appeared able to overcome their repugnance at the exterior appearance of the addition to see the merits of the dramatically sweeping interior spaces, the huge fluid lobby, and the overwhelmingly impressive new galleries for display of the museum's famous collection of dinosaur skeletons. Any of the addition's virtues seem to have been almost completely swamped by the response to the jagged edges of its exterior volume. What could possibly have been responsible for such a venomous attack on Libeskind's work?

A long tradition of research in psychology and aesthetics has established that we have a near universal preference for curved contours. This preference has been shown in many different kinds of materials ranging from simple shapes—even typography—to architectural interiors. We see curves as soft, inviting, and beautiful whereas jagged edges are hard, repulsive, and may signal risk. The contrasts between our responses to these two types of contours suggests that even low-level perceptual properties of the built environment can elicit strong responses by triggering ancient circuits that have evolved to warn us of environmental risks. Work by University of Toronto neuroscientist Oshin Vartanian has shown that exposure to curved or jagged contours in architectural interiors can change our patterns of brain activity.[8] The presentation of curves produces strong activation in brain areas like the orbitofrontal cortex and cingulate cortex—areas of our brain that are associated with reward and pleasure. Jagged edges can cause increases in activity of the amygdala—an important part of our fear-detecting and response systems. In architectural terms, it may well have been the graceful curves of Frank Gehry's Guggenheim Bilbao that reduced famed fellow architect Philip Johnson to tears when he first laid eyes upon the building. In contrast, the strong reactions elicited by Libeskind's work in Toronto, notably similar to those brought about by architect I.M. Pei's design for the new entryway to the Louvre, may have had their origin in patterns of brain responses related to our innate need to recognize fearsome environmental risks.

Straight lines and acute angles are not only less preferred and less likely to be judged as being beautiful, but living among them may also unleash potent effects on our behavior. An international team of researchers from Humboldt University in Berlin and the University of Haifa conducted an experiment on social judgment in which participants were asked to assemble a portrait of a stranger from puzzle pieces that had either rounded or jagged edges. Participants judged those faces constructed from jagged edges to be colder and more aggressive than those constructed from rounded edges. In a telling follow-up study, participants were asked to participate in a common game of economic decision-making, where they could choose to act either cooperatively to share winnings with a partner or aggressively and take all of the spoils for themselves. Participants played the game in one of two different rooms, each decorated with a collage of abstract shapes on the walls: one room sporting sharp, angular shapes while the other room displayed shapes with curves. Participants were significantly more likely to behave aggressively when they were surrounded by art with sharp angled shapes than when they were in the room where more rounded contoured art was hung. Collectively, these experiments suggest that the shapes of contours that surround us can make us feel either happy and comfortable or anxious and fearful, but they can also affect how we treat others. Such effects seem to run deep. Three-year-old toddlers show similar preferences for curves over angles, suggesting that our responses to geometric shapes arise early in life and may not depend on adult experience or matters of taste.[9]

In evolutionary terms, it makes perfect sense for us to have an aversion to hard, sharp edges and acute angles. Such shapes may suggest teeth, claws, or other kinds of dangerous edges; it would be adaptive for us to veer away from them and toward gentler surfaces. The evidence that exposure to such shapes may spill over from simple preference to more complex types of behavior such as social judgment and cooperative group behavior is also in keeping with modern views of embodied cognition in which factors such as temperature and brightness appear to exert an effect on our views of human relationships and our inclination to prosocial and ethical behavior.

Fear of the Other

Beyond obvious factors such as shape, noise, sensory overload, and exposure to environmental threats, the single factor that stands above all others as a potential stimulus for anxiety in modern built environments is the interpersonal one. Put simply, modern urban life requires us to live in close proximity with strangers. Considered across the broad sweep of evolutionary time, this state of affairs is entirely unnatural for us. Our biological heritage, and the style of living for which much of our mental machinery is adapted, involved living among small groups of individuals, mostly kin, and in groups of fewer than one hundred people whose appearance, personalities, and habits we knew well. To the extent that our ancient forebears lived among built constructions at all, they were more likely to consist of rudimentary sheds that could be used to store their possessions than as shelters that they could use to hide from one another. The shift from a lifestyle of frank exposure to the gaze of kin to the norm of modern urban living in which we live in close quarters with teeming multitudes of people about whom we may know nothing at all has been jarringly rapid. Indeed, it is this state of perpetual life under the gaze of the stranger that is probably responsible for much of the social stress identified by Meyer-Landsberg's group. Before we lived among strangers, modern conceptions of trust and privacy were radically different from how we currently view them, and probably almost superfluous. Small, agrarian groups who lived most of their lives within full sight of one another would have been much less vested in their own privacy, and even the idea of an inner life separate from the life of the group would likely have been relatively unfamiliar to them. Because of this, trust—a state in which we believe we are familiar enough with the inner state of others to rely on them to behave in a predictable manner—would also have had much less meaning than it does for modern humans who live most of their lives behind walls, both physical and metaphorical ones.

Most recent discussions of the manner in which the built setting can influence interpersonal relationships have focused on how elements of physical design might influence our feelings of togetherness, and accordingly affect many other aspects of our behavior, ranging from our desire to inhabit open, public spaces to our willingness to go out of our way to help others. Many have focused on the importance of our feelings of ownership or territoriality for spaces. The American architect Oscar Newman claimed that features of its physical design that failed to promote such feelings of ownership brought about the famous catastrophic failure of the Pruitt-Igoe Building—a major public housing complex built in St. Louis to provide shelter for the dispossessed with a particular focus on single-parent homes headed by women. Almost from the time that it was built, Pruitt-Igoe fell into disrepair and became a serious center of criminal activity. Newman argued that one of the main reasons for this had to do with the design of its shared spaces which sat empty, uncared-for, and unoccupied because the residents did not feel a sense of ownership for these spaces. Ultimately, and with considerable public attention, Pruitt-Igoe was written off as a failure and razed to the ground. More recent analyses of the failure of Pruitt-Igoe have suggested that it had at least as much to do with economics, politics, and racism as it did with architectural design;[10] nevertheless, Newman's analysis helped set into motion a major new way of thinking about how to promote secure and trusting environments through the placement of walls, doorways, and public spaces in housing developments. His landmark publication, *Defensible Space*, outlined a series of straightforward architectural design features that he argued could have saved Pruitt-Igoe by promoting feelings of community and shared vigilance among its residents.

In less dramatic fashion, other studies have shown how the layout of a building interior or neighborhood can influence our feelings about the strangers who surround us, and thereby our behavior toward them. In one such study, experimenters tested prosocial behavior using a clever method that involved scattering stamped, addressed letters on the ground and using as a measure of prosocial behavior the likelihood that such letters would be picked up and mailed by strangers. They conducted their experiment in a series of university dormitory buildings with dramatically different design features. High-rises housed large numbers of students who seldom crossed paths in the course of an average day. Medium and low-density buildings, such as low-rises or townhouses, had more shared facilities, including dining facilities where students typically saw one another at least once during the day. The researchers found that the rate of return of dropped letters was strongly influenced by the physical layout of the building in which the letters were dropped, with the highest rates of return in the low-density buildings (in fact the rate of return in such buildings was 100 percent!) and the lowest rate of return in the high-rises (just over 60 percent of letters were returned in such settings).[11] Related research has shown that students who live in the lower floors of high-rise dormitories are likely to have richer social networks within their residences than those who live on the upper floors, presumably because residents who live on the lower floors are more likely to use the common areas that are found at ground-level in such buildings, and so will be increasingly likely to recognize their co-residents. Lower floor residents also seem to have increased levels of trust in their neighbors and are happier with their living arrangements than those who live at the lofty peaks.

What these studies and many similar ones suggest is that the way that our built environment is organized, and the manner in which it influences our day-to-day interactions with other people, will have a tangible effect on our feelings of trust, our willingness to help strangers, and our happiness with our living conditions. This research contains valuable lessons for urban designers and architects who might desire to influence our sociality, but it also points the way toward a deeper understanding of the relationship between place and anxiety. To see the connection, we need simply ask what it is about environmental conditions in which we find ourselves in a crowd of strangers that *prevents* us from behaving in a friendly and prosocial manner. There are many possible answers to this question, but most of them boil down to one simple idea: we keep safe distance from strangers because we fear harm. This conclusion may seem extreme. Although most of us behave with some level of reserve among strangers, it might seem a gross overstatement to suggest that the source of this reserve is the fear that if we walk up to a stranger, smile, and begin a conversation, there is a tangible risk that he will hurt us. Yet it *is* plausible that the main reason why we might not greet a stranger might be a fear of exposing our private inner selves to the scrutiny of an unknown outsider, and one could argue that this is simply a different kind of risk, and one that we might be biologic-ally disposed to avoid. Living in crowded conditions among strangers unleashes a large spectrum of protective impulses that can range from our simple reticence to engage in conversation with the person who sits beside us on a city bus to our reluctance to take a walk through our own neighborhood after dark. All of these responses are, in one way or another, adaptations to unnatural living circumstances that alert us to the presence of risk and so provoke some level of fear or anxiety.

Fear of Crime

In 1969, Stanford social psychologist Philip Zimbardo conducted a simple and daring experiment. He parked cars in two different locations: one in a sketchy neighborhood in New York's Bronx and the other in Palo Alto, California, near his home university. The license plates were removed from the cars and the hoods were raised to suggest that the cars had been left following an episode of mechanical trouble. Zimbardo's research assistants waited nearby but out of sight to watch and film the result. In the Bronx, the abandoned car was stripped quickly. The acts of vandalism began almost before the assistants had a chance to move out of sight and to set up their camera. In Palo Alto, the car was left intact for many days. Indeed, one passerby lowered the hood of the car during a rainstorm to protect the interior. Zimbardo interpreted this straightforward result as having been a result of differences in feelings of community and reciprocity in the two neighborhoods. Just as the hallways of Pruitt-Igoe had apparently belonged to nobody, the streets of the Bronx were not considered to be a part of the shared space of a community with its inherent requirement that residents watch over and care for the contents of the space.

In a second phase of the experiment, Zimbardo took one additional step: he smashed the windshield of the car in Palo Alto. Not long afterward, he began to see the same acts of theft and vandalism toward the car at the second site as he had seen in the Bronx. Political scientist James Wilson and criminologist George Kelling used this simple observation, publicized not long after the experiment in an article in *Time* magazine, as the cornerstone of a major new theory describing the origins of urban crime. The key argument of Wilson and Kelling's so-called broken windows theory was that physical signs of disorder—broken or boarded up windows, litter, or graffiti—served as overt signals that nobody cared about the surrounding environment and this evident lack of caring encouraged crime. If Wilson and Kelling were right then a key corollary would be that any efforts taken to minimize signs of physical disorder would also discourage crime. The theory, with its straight-forward prescription for crime reduction was taken up with gusto by city officials—first by William Bratton, the head of security for public transit systems in New York City, who made great efforts to sanitize its subway system, and later by New York Mayor Rudy Giuliani, who with help from Bratton—by now New York's police commissioner—extended the definition of disorder to include *social* disorder, the perpetration of petty crimes such as public drunkenness and urination, fare-dodging and panhandling. A major crackdown on both physical and social disorder followed, including controversial enforcement measures that allowed police to stop, question, and frisk any suspicious individuals appearing on the streets. In New York and a few other places in the United States and beyond, a measurable drop in serious crime followed these initiatives; so the application of broken windows theory to the problem of urban crime was lauded as a success.[12]

Research and debate about the broken windows theory continues unabated and with some heat. Critics argue that the decreases in crime rate in New York observed during Mayor Giuliani's reign also coincided with a general increase in the standard of living of residents of the city and notably a precipitous drop in the unemployment rate. Economic factors such as these would also very likely exert downward pressure on many different kinds of crime. Spatial analyses of crime statistics have suggested that the fit between careful measurements of disorder and observed crime rates, though sometimes generally supported, is not as compelling as might be suggested by Wilson and Kelling's theory, and that measures designed to "sanitize" neighborhoods of disorder are dis-criminatory against the poor and the dispossessed. But regardless of whether crime rates are strongly predicted by measures of disorder or whether reducing disorder actually does reduce crime, there's a second focus of interest on the influence of disordered environments. This second focus, though less often squarely in the sights of criminologists, probably has much more to do with the quality of our emotional lives in cities. This second factor is our *fear* of crime.

In a survey of fear of crime in a broad sampling of European countries, about one-third of respondents reported that they sometimes altered their behavior—avoided parts of the city, altered planned walking routes, changed the timing of the events of their lives—out of a concern for personal safety.[13] North American studies show similar patterns. In one large U.S. study, fully 40 percent of respondents said that they would be fearful about taking a walk at night within a mile of their homes. Even when inside their homes, a substantial fraction of survey participants (particularly women) reported some level of concern about home invasion.[14] These numbers are in striking contrast to the annual statistics on actual reported crimes, which even in the direst cases hover between 1 and 2 percent as a proportion of a country's population. In addition, there is often only a weak relationship between the incidence of real crime and our fear of crime. For example, although the crime rate in Sweden is among the highest of all western European and Nordic countries, fear of crime in this country, along with other Nordic states, is extremely low. To some extent, there are probably complicated sociological causes for these disconnects between our fears about our environment and its true level of risk. Our fears of the worst crimes, such as homicide, are the ones that are most seriously out of kilter with realities, but this is likely because such crimes are the ones that are most likely to appear in media reports. Our general level of fear is mediated both by our feelings of community and connectedness to our neighbors and by our trust in police forces, and these can be greatly influenced by social and governmental policies. But notwithstanding these mediating effects, one might argue that our fear of crime, because it is so high compared to the real incidence of crime, is maladaptive because it interferes with the normal conduct of our daily lives. But considered in light of the bald probabilities of crime, the risks of personal threat, and the costs of mitigating these risks by alterations in our personal behavior, the matter is less clear.

In an early part of my career, I studied animal behavior. How do animals arrange their lives to fulfill their ultimate biological mandate to reproduce, thus ensuring that copies of their genes are carried into future generations? My main focus was on the manner in which prey animals take measures to avoid being killed by predators. Much of this work was conducted in laboratory environments in which I built elaborate contraptions such as flying cardboard hawks that were meant to induce laboratory animals like rats, mice, and gerbils to run for their lives and seek shelter. But for some first-hand experience of prey–predator interactions, I used my professional interests as an excuse for an adventure in Kenya (which my department chairperson, with considerable acumen, referred to as "Ellard's monkey business"). Based on my knowledge of African wildlife at the time, most of which had come from watching Alan Root's famous nature documentaries, I expected to see dramatic chases between predatory cats and bounding gazelles. As I crashed around Masai Mara in the back of a jeep driven by my guide, who had a terrific knack for stranding us in the middle of irritated herds of wildebeest or elephants, I had the naïve expectation that I would see skulking cats creeping toward unwary prey animals, all culminating in magnificent chases. I had the dreamy idea that somehow I might discern what separated a successful evasion from an unsuccessful one. What is the magic formula for avoiding being eaten? It didn't take long at all for me to have my first and most important lesson in prey–predator interaction. It isn't a game of stealth and chase so much as one of dreary economics. On the wide open terrain of the savannah, the most common type of interaction between prey and predator is a kind of calm standoff—a game of carefully calculated costs and benefits. The gazelles, standing and grazing in a herd, can see a lone cheetah. The cheetah watches the herd. Both know exactly what's going on, but the gazelles don't stampede over the nearest hill in panic. They want to keep eating for as long as they can. Flight distance—the distance over which the cheetah could close the gap between it and the herd to bring down a victim—is carefully calculated and recalculated. As long as the gazelles remain outside that distance, they are reasonably safe and can continue to forage. The lesson here is that there is a *cost* associated with unnecessary alarm and flight reactions, and prey animals are adapted in such a way as to only bear that cost when it makes sense to do so. The long-term outcome of the game depends in large measure on the accurate calculation of such costs.

Let's come back from the plains of sub-Saharan Africa and consider life in the city. Whenever we make a decision based on our feelings of anxiety—inner warning bells that we may be placing ourselves at risk—we are behaving in much the same way as a grazing gazelle in Masai Mara. We may choose to drive rather than to walk at night or we may cancel a plan altogether. We may take a longer route to avoid an area that we perceive as risky. We may cross the street or turn around and backtrack to avoid close contact with a group of people loitering on the street. Such choices are economic decisions based on our weighting of the relative benefits of doing what we really want to do and the calculated risk that harm may befall us if we do. When we are behaving adaptively, the calculations of these risks and benefits are the things that we actually *feel* in our bodies as our level of anxiety. Given that we seem to generally overestimate risks, one might argue that we are thwarting our desires unnecessarily, but based on gazelle thinking, it might simply be the case that the overall costs of choosing behaviors that appear to be safer are very low compared to the costs of discounting our feelings of anxiety too heavily. Evolutionary psychologist Robert Ornstein lays out the mathematics like this:

> Fail to respond to a real danger, even if that danger would kill you only 1/10,000 as often, and you will be dead. A few years later, you will be deader in evolutionary terms, for fewer of your genes will be around. However, an overreaction to danger produces only a little hysteria . . . no loss of reproductive ability. . . . If panic in response to a threat in all cases improved survival by even 1/10,000, those who

panicked would be 484 million times more populous than those who did not.[15]

Just as there may be evolutionary advantages to behaving like an overly twitchy gazelle, we also seem to have some straightforward environmental triggers for arousal, anxiety, fear, and avoidance, most of which are likely of ancient origin. In empirical studies of fear in urban settings, the most important triggers of feelings of risk are related to spatial properties. We don't like walking into situations where potential flight paths are blocked, we don't like walking through areas that contain lots of shadowy hiding places for potential lurking no-goodniks, we don't like walking toward an area where it's difficult to see what's around the corner, and we don't like walking through areas that are completely empty of people. In some cases, cues of physical or social disorder can heighten anxiety; of course, our knowledge of an area, developed either through personal experience or through media accounts of violence, will also dissuade us from venturing into territory that may be unsafe. There are also some perfectly sensible contextual and individual variables that can influence our feelings of personal risk. We are much more likely to be circumspect at night than we are during the daytime. Women and the elderly have lower thresholds for anxiety or avoidance, and this is perfectly in keeping with their greater vulnerability to threat.

The gender difference in both perception of and vulnerability to risk is difficult to overemphasize and should be a key element of successful urban planning. In 1991, A Viennese survey found that the daily routes of men and women through the city was markedly disparate: men tended to drive or take public transit twice a day, once on the way to work and once on the way home again, whereas women took varied routes related to childcare, household shopping, and a variety of other activities. In response, Vienna instituted a policy of "gender mainstreaming" designed to promote equal access and opportunity for both men and women in urban environments.[16] Some elements of this policy—improvements in lighting and the design of walkways—were explicitly designed to address gender differences in both fear of crime and victimization. Vienna's policies, now overseen by a government department and an action plan supported by budgetary principles that require consideration of gender issues in many aspects of city planning, should act as a model for any city working to eliminate gender disparity of access to public places and amenities.

Protecting the Inner Self

In some ways, anxiety related to fear of bodily harm in built settings is a straightforward thing. We understand the tangible risks to our bodies and our possessions at the hands of criminals, and though some might argue that our emotional responses to those risks are out of kilter with the realities, most of us have a well-tuned set of sensitivities to the environmental factors that suggest risky environments—at least some of them having developed through thousands of years of evolutionary adaptations—and we respond to those sensitivities with adjustments to our behavior.

Compared to such responses, the behaviors that we undertake to protect our inner selves from exposure to the world of strangers might seem somewhat more ephemeral because unlike in the case of avoidance of crime, the stakes are not always so easy to define. Although it's certainly possible that revealing our inner secrets risks both material and psychological harm (think of identity theft or psychological hazing through electronic social media like Facebook or Twitter for obvious examples), the measures we take to protect personal privacy extend far beyond considerations of these kinds of risks. Striking up a conversation with a stranger on a bus can require us to overcome a threshold of reserve, but it's very unlikely that doing so will place us in a position of risk. For urban dwellers, there is also some irony in the fact that the same set of impulses that we use to protect ourselves from the prying eyes of strangers is partly responsible for one of the great psychological blights of big cities: the epidemic of loneliness.

In many parts of the world, we are in the midst of an important demographic shift. In 2013, the proportion of single people in the U.S. population rose above 50 percent for the first time since accurate records have been kept, representing a dramatic shift that has taken place in a relatively short period (the proportion of unmarried individuals has risen by about 35 percent since 1976).[17] Similar shifts have taken place in northern Europe where, for example, in London the proportion of single member households topped 50 percent in 2011, and other regions of the country experienced precipitous increases in the number of singleton households.[18] This change has enormous implications not only for our politics and our culture, but also for the manner in which we use every kind of space from our homes to our public places. At the same time, some sociological studies have suggested that our circles of intimate friends may be shrinking steadily. In one study, for example, a large sample of people in the United States was asked to list the people with whom they could discuss "important matters." The average number of close confidantes in this sample was calculated as 2.08: a drop of about one person compared to a similar study conducted about a decade before this.[19] Considering that one's closest confidante might ordinarily be supposed to be one's spouse, this decrease may not be surprising. During the same period, there has been a precipitous rise in our use of electronic social media. A quick Google of the word *friend* shows that many of the top hits have nothing to do with things like getting a cup of coffee or a glass of wine with a companion to share the events of the day and more to do with setting privacy levels correctly on Facebook.

Collectively, these changes—drops in marriage rates, increases in solo living, and the ubiquity of online social networks—have unleashed a torrent of studies, books, and opinion pieces on their implications for human social behavior. The dust hasn't even begun to settle on matters yet, but certain things seem clear. In a recent survey in the city of Vancouver, study participants reported that loneliness was their most significant issue with city life, rated ahead of any other economic or lifestyle issue.[20] A similar study in Australia showed that the percentages of respondents who felt that they had no friends at all had risen by 50 percent between 1985 and 2005. The same study revealed that fully 13 percent of respondents felt they had nobody in their neighborhood that they could turn to for help if they needed it.[21] Although as psychologist John Cacioppo points out in his landmark book *Loneliness: Human Nature and the Need for Social Connection,*[22] being alone is not at all the same thing as being lonely, it is certainly clear that low marriage rates, high levels of solo living, and generally smaller social groups pose challenges for people who are trying to avoid the punishing psychological state of loneliness. Much research has shown that loneliness exacts high costs. Those of us who experience it chronically are likely to suffer from depression, low self-esteem, diminished opportunities for advancement, and even disease and early death. The factors responsible for the surge in loneliness are not completely clear, though some have pointed the finger at the trend to decentralized suburban living and its consequent long commutes to workplaces. Others have suggested that electronic media, especially the Internet and online social networks, may have made it possible for us to live functionally, though not necessarily happily, in greater isolation from our fellow human beings. We can shop, play, and even gather into social groups (after a fashion) without leaving our homes. Although it does seem to be the case that our social groups are becoming smaller and more polarized, the relationship between these shifts and the pervasive presence of online networks is considerably less clear. Indeed, some of the best evidence, coming from a study by sociologist Keith Hampton, suggests that although we may name fewer close confidantes than ever before, a part of the reason for this may have to do with our changing definitions of friendships and the manner in which we organize our social lives.[23] Further, Hampton argues that there is little evidence that these changes have been driven by the use of online networks. Some studies have even shown a direct correlation between the sizes of our online networks and our real-life social networks and a further correlation with the sizes of some brain areas—such as the amygdala—thought to be involved in the regulation of some aspects of our social lives.[24] In addition, some studies have shown that online networks designed to provide information to local neighborhoods can exert a dramatic positive impact on neighborhood social cohesion.[25]

Despite the haze of confusion that surrounds the interrelationships of social networks—both real and virtual—and the design of built spaces, one thing seems clear: considered over the broad reach of time, one of the most important changes that has taken place as we have moved from small group living to cosmopolitan lifestyles in large cities has been that we have lost the ability to be familiar with every person we see on a daily basis. The swarm of individuals we live among has vastly exceeded our cognitive capacity to keep track of one another. In his book *Eavesdropping: An intimate history,*[26] John Locke has argued that this transition was a pivotal factor in human social development because prior to this, basic human needs that we now deem precious to us such as privacy and trust had much less significance. When everyone is visible almost all of the time, the very idea of an inner self seems exotic. Considered in this light, it might seem less strange that our modern condition prompts us to initiate overly zealous protective responses by provoking anxiety responses when our private inner selves are at risk of exposure. One could consider life in urban conditions to represent an uneasy compromise between behavioral patterns that have evolved to deal effectively with life in small, exposed groups and a physical environment that forces us to rub shoulders with thousands of strangers. However, unlike the case of anxiety responses that may protect us from crime, where one could justify a low set-point on the basis of a cost–benefit analysis that suggests that it's better to err on the side of a "better safe than sorry" strategy, the costs of social isolation provoked by life in a crowd of strangers may be somewhat higher. By insulating ourselves from the stranger with whom we share an elevator or the next person in a queue at the supermarket, we may not only be depriving ourselves of pleasant acquaintance with others who are just like ourselves, but we may also be driving our anxiety levels and our stress responses to unhealthy high values. This is certainly the type of thinking that encourages urbanists like Charles Montgomery in his book *Happy City* to advocate for planning and design changes in cities that encourage affiliative behavior—amenable public places, pleasant urban green areas, and domestic living arrangements such as low-rise condominium developments that help to push us toward one another in circumstances where our moods are likely to be high and positive.[27]

The same kind of dynamic may also account for the rampant popularity of social networking sites such as Facebook. One could consider the always-on nature of status updates on Facebook newsfeeds to represent the modern equivalent of ancient rituals whereby small groups of agrarian settlers monitored one another's movements visually as they sat around a small circle of hearth spaces, always in plain sight, but not necessarily in direct communication with one another. The flickering stream of status updates, always available when you want to look, but not necessarily demanding focal attention, strike a strong parallel with these kinds of primitive, casual exchanges of information. It's interesting in this regard that the median number of Facebook friends for a user is around two hundred—within close range of the famed Dunbar number of 150, argued by anthropologist Robin Dunbar to be the rough capacity of the human mind for stable relationships and a close match with the average number of participants in many different kinds of social organizations ranging from Neolithic farming villages to modern military company sizes.[28]

Although the comparison between our use of online social networks for monitoring of ubiquitous social presence and the behavior of Neolithic farmers gazing around the fireplace might be tantalizing, there is one very important difference between these two types of networks. Our natural, organic networks are bottom-up and self-organizing. We monitor one another's actions and thoughts using simple metrics based on gaze and hearing. Somehow, our mutual actions and observations generate a kind of group understanding and cohesion. In the case of an online network like Facebook, it's very clear that in addition to our own contributions to the network through the posting of words and images, there is a layer of executive control. The Facebook organization monitors our contributions, filters them according to their own closely titrated algorithms, and is even known to experiment on us occasionally to tweak the network and presumably maximize its penetration into our daily lives for the purposes of profit.[29] This layer of control, more or less invisible to the casual user except for the presence of context-sensitive advertising, makes the use of proprietary social networking sites an entirely different kind of thing to what happens around a camp-fire, and potentially quite a bit more worrisome. Our compulsion to use these networks, perhaps born of a yearning for the kind of perpetual social monitoring that worked well for small groups of early humans, may represent a response to the anxieties and fears of detachment that arose when we began to move into large cities.

CHAPTER 6

PLACES OF AWE

O N CHRISTMAS EVE, 1968, Apollo 8 astronaut William Anders took a photograph that was destined to become one of the most famous images in human history. As the tiny spacecraft that he shared with astronauts Frank Borman and Jim Lovell rounded the moon and revealed the blue globe of planet Earth, Anders raised a Hasselblad camera, exclaiming with all the enthusiasm one is likely to ever hear from a fighter pilot with the United States Air Force: "There's the Earth coming up. Wow is that pretty."[1]

As a young boy, I remember the excitement of plastering a poster of this image on my bedroom wall and looking at it longingly every night as I went to bed. My thoughts about it were of the simple ten-year-old kind. To me, it meant adventure, exoticism, and remoteness. It enlivened in me what most young boys would have felt at the sight of it: a desire to explore, a feeling that now that human beings had escaped the planetary atmosphere in earnest, anything at all was possible and we would all soon be living on starships. The future was filled with untold possibilities. Over the years, I've seen this image everywhere, from countless museum exhibitions, and school science fairs, to cheesy montages of images on the walls of souvenir shops, and even at a tiny drink stand on a scalding piece of dirt road between Nairobi and Mombasa. We gravitate to it not only because of its intrinsic striking beauty, but also because of what it helps us to remember: we are all inhabitants of a tiny planetary spacecraft that is hurtling through an essentially limitless expanse of space. Writing in the *New York Times* on Christmas Day of 1968, the day that the image began to circulate in earnest, the poet Archibald MacLeish captured the feelings of many by saying: "To see the Earth as it truly is, small and blue and beautiful in that eternal silence where it floats, is to see ourselves as riders on the Earth together, brothers on that bright loveliness in the eternal cold—brothers who know now they are truly brothers."[2]

The feelings experienced by the three astronauts on board Apollo 8, and those enjoyed vicariously by those of us who watched in wonder from Earth's surface, have been felt over and over again by astronauts venturing into the far reaches of space. Author Frank White dubbed the effect that I am describing as the "Overview Effect" in his 1987 book, which took the name of the effect as its title. In the short film *Overview* from White's group, the Planetary Collective, philosopher and Zen teacher David Loy describes the feelings of astronauts gazing at the Earth as a realization of "their interconnectedness with that beautiful blue-green ball." He further defines the near universal response of the astronauts as a feeling of awe, which he defines in part as a willingness to "let go of oneself. To transcend the feeling of separation."[3]

Although very few of us have been lucky enough to travel into space and experience awe by looking at the Earth from a remote viewpoint, everyone has had experiences that they would categorize as "awesome" (and not just in the recent banal sense of that word). When awe strikes us, we are certain of it. We can be overcome by awe when we encounter a dramatic natural phenomenon such as an inky starlit sky, a thunderstorm, or a majestic view of a mountain range or canyon, or even by simple reflection or discussion of major world events (for example, simply hearing the radio transmissions of the conversations of lunar astronauts generated a massive, global experience of awe). More to the point in the current chapter, we can also be overcome by awe in built settings. But what, exactly, does this mean? Dry, dictionary definitions suggest that awe consists of a unique combination of surprise and fear. But the reflections of astronauts who experienced the Overview Effect suggest that feelings of awe also include elements of transcendence. Such experiences bring us outside the narrow confines of the body space, encouraging us to believe that our existence constitutes more than just a beating heart inside a fragile organic shell. We have a sense of boundlessness as the limitations of time and space that hold us aground are suddenly swept aside.

Much of our interaction with place is driven by deep biological imperatives that we share with almost all other animals. Just like us, other animals can benefit from (and perhaps even enjoy) the feelings of enclosure and comfort that can only be found in a home space; we have no strong reason to believe that the experience of a young rabbit huddled in a burrow is markedly dissimilar to that of a child at rest in her bedroom. We experience the pull of our basic attractions to novel sights and sounds in a casino or shopping mall, and we may be driven to spend beyond our means. Though a laboratory rat could not describe a rich phenomenological experience during a frantic session of bar-pressing in a Skinner box in hopes of a rewarding squirt of chocolate milk, the structure of the behavior, and even its neural underpinning, is not significantly different from our own. Indeed, it's very likely that even the feelings we experience during such occasions, feelings of craving and want, for example, are not qualitatively different from those felt by other animals. We feel anxious when we detect threatening circumstances, and it is perfectly straightforward to see the connections between those feelings—and the actions that might stem from them—and the evolutionary roots of our reflexes to avoid being someone else's dinner. But now we turn to a type of reaction to place that, for all we know, may be uniquely human: the experience of awe that can be generated by immersion in transcendent places.

Although early psychologists such as William James (who sometimes to me seems to have thought of everything important in psychology before anyone else had a chance) certainly discussed awe, and even psychodynamic theorists such as Sigmund Freud, Carl Jung, and Otto Rank described an important role for the transcendent experience in their theories of human behavior, more hard-nosed psychologists have come to the table only fairly recently. There are many reasons for this, some related to the fact that awe has never really been considered as a primary emotion on the same level as emotions such as fear, surprise, and disgust. Yet this too may be related to the human uniqueness of feelings of awe. Again, we have no trouble thinking about the existence of primary emotions in other animals—and it is very clear that animals experience such emotions—but a part of the reason for this has to do with the clear sense in which such emotions are adaptive. Whether you are a human being or an armadillo, it makes sense to defend when afraid, run when surprised, and avoid when disgusted. But what is this uniquely human emotion of awe for? How does it work? What, exactly, causes it? How is it measured?

Following an incisive analysis by psychologists Dacher Keltner and Jonathan Haidt of the manner in which the word *awe* has been used by theologians, sociologists, psychologists, and lay people, many current researchers focus on a pair of distinctive properties that seem to be shared by all awe experiences: a feeling of vastness and a sense of accommodation.[4] Vastness may be experienced physically in the way that we might feel while peering over the edge of the Grand Canyon, but it can also be experienced indirectly. Superheros might generate a feeling of vastness by virtue of their enormous range of special powers. A brilliant intellect—an Einstein—might generate awe because of great powers of understanding. But underlying all feelings of vastness is a sense of the "bigness" of something.

Accommodation describes the manner in which we may be required to adjust our world-view in response to the stimulus that generates awe. Related to the epiphanic experience, the crux of accommodation is that it normally brings together two things—ideas, notions, or even sensory experiences—that are contradictory. The only way to overcome our feeling of contradiction, the "at-oddness" of the experience, is to adjust what we thought we knew of the world, and sometimes on a grand scale. A good scientific example would be the mental struggle of a student of physics learning for the first time that light can be described as both a wave and a particle. Not coincidentally, many examples of accommodation and epiphany relate to religious experiences. In Christianity, the notion that Jesus was both a mortal human being (as amply demonstrated by his murder) and a god (as shown by his return from the dead), is a sublime example of a contradiction that requires an active reconfiguration of one's world-view (or perhaps more aptly one's universe-view) in the form of a supreme act of accommodation to make sense of such a duality. This was certainly true of the accounts of astronauts experiencing the Overview Effect, who came to see their interconnectedness with the universe on a grand scale that required an act of accommodation, and it is a feeling that many of us have had in other kinds of circumstances as well.

When I entered St. Peter's Basilica in Rome for the first time, I was overcome by both the vastness of the interior and the sheer quantity of ornate decoration and artistic riches that it contained. My initial impression of the space was amplified by the bodily reactions I observed in other visitors. Religious pilgrims slumped to the ground and began to crawl from the narthex of the building to its transept. Even casual tourists looked as though they had had the wind knocked out of them. My own epiphanic experiences centered in part on a new understanding of the power of a built structure to effect such a strong feeling regardless of one's own religious beliefs (which in my own case I could only describe as scanty). Indeed, this experience was an integral part of the development of my current interest in the power of built structure to organize thought and feeling. Intellectually, the idea that place mattered was one that I was well familiar with, but there is an enormous difference between understanding the minutiae of technical accounts of our psychology and standing in an enormous basilica and being bludgeoned by an undeniable force. In addition to the feelings of vastness and the accommodative reaction related to my intellectual pursuits, I also felt a sense of union not only with my fellow visitors to the site, but to all of those who had come before me in the centuries preceding my visit. Just as astronauts report having felt a melting away of time and space, a rupture of the discrete borders that separated their own bodies from the rest of the universe, I felt a similar breaking down at the edges of my selfhood and a similar sense of mystical union, made all the more remarkable by the fact that I knew that what I was feeling was, in some sense, *intentional*—that one of the purposes of the building in which I

stood was to deliberately elicit what I was feeling and to use those feelings to change who I was.

Although it might be true that the full-blown experience of awe
that I have described, the sense of vastness and accommodation along
with a feeling of mystical union to a greater thing, is a uniquely human
experience, the beginnings of awe, its evolutionary antecedents in the
world of nonhuman animals, is more prosaic and completely in line
with some of the other human emotional capacities that I've described,
and vastness is the real key to the experience. Anyone who has ever
seen a small dog cowed into submission by a larger beast—the bully of
the street—understands the power of vastness. The small dog is carrying out a script that on the face of it seems motivated by an interest in
self-preservation. If it doesn't roll over and expose its belly, we know
that the conflict is much more likely to escalate and that the larger dog
is likely to accelerate its attack, possibly resulting in serious injury. The
more interesting question is why the conflict arises in the first place.
Why don't the two dogs simply ignore one another and walk on by?
A complete answer to this question would take us far from our main
topic, but a satisfactory answer is easy enough to sketch out. Living in
groups conveys certain advantages. Groups of animals are better able
to defend themselves from attack and just as importantly, they are able
to defend the resources that accrue to their territory. But as anyone
who has shared a dormitory room or an apartment with a group of
other individuals knows, such sharing comes with a cost: some of us
will be able to use our might or our wiles to obtain a bigger portion of
those resources. To address this dynamic, one alternative would be to
fight to the point of severe injury or death every time a resource conflict arises, but as animal behaviorist Konrad Lorenz first pointed out
in his landmark book *On Aggression*, it makes better adaptive sense
for animals in conflict to solve their problems by means of signaling
to one another the likely outcomes of battle so that it doesn't need to
take place in the first place.[5] For our poor little schnauzer out for a
Sunday stroll, contemplating the outcome of combat with the spike-
collared bulldog it encounters, this means one simple thing: the little

dog knows that it will lose so it gives up before the battle starts. This simple piece of survival algebra accounts for a great deal of the social behavior that takes place among groups of animals, including human beings. The advantages of size are obvious everywhere. In nonhuman primates, the largest and most powerful males are the ones who enjoy privileged access to food, shelter, and mates. In human beings, the tallest men make more money and enjoy greater social status. We even honor size symbolically by inscribing the names of more powerful individuals in larger fonts or by placing them in the highest offices in a building. And although we might not like to think of ourselves as taking part in daily duels with larger and more fearsome members of our species for possession of scarce resources, it seems that this understanding of the power of size in social relationships is something that is instilled in us before we can even speak.

In one clever experiment conducted by a team of psychologists from Harvard University led by Susan Carey, preverbal infants were shown movies of simulated "combat" between two simple animated squares—one large and the other small—possessing eyespots and mouths.[6] The squares acted as though they were trying to pass one another on an avenue (very much like our two hypothetical dogs), but there wasn't room for them to pass and so a shoving match ensued. In one condition, the smaller square engaged in a kind of a submission display in which it slung itself close to the ground, allowing the larger square to pass. In the other condition, the larger square submitted. As the infants watched these displays, the researchers carefully measured their gaze. How much attention did they pay to the simulated displays of dominance and submission and were there any differences between the two cases? Remarkably, by eleven months of age, infants were much more interested in the displays that showed the smaller square dominating the larger square, suggesting that the infants sensed the novelty of a situation in which the normally expected dominance hierarchy between the squares was reversed. In other words, before they can use language, human infants have inbuilt mechanisms for understanding the normal operation of social dominance.

We've veered some distance from a consideration of the adaptiveness of awe, but there is a strong connection between this near universal observation of the importance of size and the nature of our responses to vastness, especially the kind of physical vastness that we are likely to encounter in a large building like a cathedral. The straightforward Darwinian argument is that our responses to large buildings and other kinds of vast objects like the Grand Canyon or an inky sky filled with stars is tapping into brain mechanisms that have evolved to preserve the rules of social order and to cap aggression among competitors by encouraging the submission of the weak.[7]

Other than our innate responses to the brute size of things, there are other lessons from nature and evolution that may pertain to our responses to religious monuments or any other large structures that we might encounter in the built environment. To understand the stakes, consider the behavior of the bowerbird. Male bowerbirds build enormous structures whose sole function is to attract mates. These bowers have some remarkable resemblances to large human-built monuments. Bowers normally include a kind of causeway that is intended as the avenue of approach for potential mates. As the female approaches the bower's enormous court, she may be entertained by the call of the male bird, amplified by the resonance chamber of the structure, thus making the male bird appear to be bigger than it is. Remarkably, John Endler, a biologist at Deakin University in Australia, has reported that male birds will construct the causeway to the bower in such a way as to produce a kind of perspective illusion.[8] By carpeting the ground in the causeway with objects whose size gradually increases as the female approaches, the male is actually reversing the normal gradient of texture seen as we view the perspective of a scene. Normally, objects that are further from the eye produce smaller images on the retina, and this is one very basic visual cue that we use to assess distance and dimension. The bowerbird's artful reversal of this gradient has the effect of making the male bird appear to be larger than it really is, presumably leaving the female suitably impressed. But beyond the sheer size of the bower, it is a complex structure that requires the male bird to divert considerable time and energy from the normal workaday business of feeding itself and defending itself from competitors and predators, and it is really the cost of building a good bower that is thought to be significant. Like the enormous tail plumage of a peacock, which has no other purpose than to demonstrate that the male is suitably endowed to carry around with it extra weight, living quarters for parasites, and stark visibility for potential predators, the bower advertises the fitness of the male by illustrating that it can survive even under the duress of an enormous self-imposed handicap. In the same way, the construction of a monumental building—an Angkor Wat, a pyramid at Giza, a giant cathedral—advertises to any who see it that its creators have the resources to build to excess. It is a frank demonstration of power.

There is little doubt that the impulse to build large, expensive structures whose size, might, and decoration far exceed their function as buildings springs in part from the same kinds of motivations that cause birds and other animals to build elaborate structures in an attempt to woo mates or that cause the largest members of a social group of animals to achieve social dominance while rarely needing to use teeth or claws to defend their right to occupy the top of a dominance hierarchy. In all such cases, the real idea is to use size and investment to demonstrate might and thereby to promote the preservation of social order. But are there other ways of thinking about the function of these majestic human structures that have less to do with matters of Realpolitik, and that perhaps might serve to differentiate what we are doing when we erect a massive cathedral from what a bowerbird does in an effort to attract a mate? In addition to their function as visible symbols of power that might encourage the more humble members of society to toe the line and contribute to societal cohesion, do such buildings have any other kinds of effects on our behavior? Given my description of my own response to St. Peter's Basilica, which included a sense of dispersion through time and space and a feeling of unity with a larger existence, it seems that there must be more at play.

To see what this other way of construing the function of a large, monumental building might be, we must turn to a different part of our psychological makeup, and a part that most people would agree constitutes the essence of what it means to be a human being—our self-consciousness.

At one point or another in time, but perhaps particularly in our earlier years when we are trying to come to terms with the greater meaning of our lives, most of us struggle to understand what it means to have self-awareness. Indeed, many people, myself included, can remember with more than vague approximation the moment at which we realized that there was something special about the magical theater of our minds that, though we could never really prove it, we suspected we shared with all other sentient human beings. We are aware of ourselves. During every waking moment, we live with a fundamental division, perhaps even a kind of a contradiction, between our inner, private, mental lives and, essentially, everything else in the universe. And although there are good reasons (namely, dolphins and elephants) to have some reservations about whether or not we are *entirely* alone in this regard, there seems little reason to doubt that when it comes to full-blown consciousness, self-awareness, and the rich phenomenology of private experience, we are probably all alone on the stage of planet Earth, and for all we know currently, the entire universe.[9]

For all of our progress in understanding how our brains produce behavior, in some cases even at the level of molecular biology and genetics, this single staggering fact of human existence has effectively repelled complete understanding for as long as humans have had, well, self-consciousness. Indeed, although the landscape has changed in recent years, the prevailing view of most neuroscientists for much of the short history of the discipline was that the problem of understanding the kernel of self-consciousness—providing a true reductionist account of what self-consciousness might even be—was simply intractable. The so-called hard problem of consciousness was thought to be beyond the ambit of the scientists' electrodes and brain-imaging machines. Even understanding what consciousness might be *for* has been a daunting problem. Although most of us might feel that something as singularly dazzling as the ability of a human being to enjoy privileged access to internal mental states, defining as it does the human condition as nothing else really can, must make some key contribution to human survival and flourishing, we've been largely at a loss to say precisely what that contribution might be. I recall once teasing a classroom full of graduate students by challenging them to think of any aspect of human behavior that could not be fully accounted for without invoking self-consciousness at all, suggesting that if none of us could define its contribution to our behavior then it might be entirely superfluous—nothing more than a fancy bauble to be marveled at, but not worried about. If it had no clear function then it might not even exist. As a philosophical position on consciousness, this idea is nothing new. For centuries, so-called epiphenomenalists have suggested, following the English biologist Thomas Huxley, that the raw contents of consciousness might have no more to do with adaptive behavior than the sound of a steam whistle has to do with the work of a locomotive engine.[10]

Before we disappear completely down the rabbit hole of difficult theoretical arguments at the interface of philosophy and neuroscience, it might be best to cut our losses for the moment and assume that consciousness is for *something*, though we may not know exactly what that something is. Perhaps, following Nicholas Humphrey's relatively straightforward argument in his book *Soul Dust*, we might simply argue that whatever else may be true, having a lively inner existence makes one's life infinitely more interesting, and perhaps *that* is really the point of it.[11] When Woody Allen's character Isaac Davis, in the movie *Manhattan*, provides a list of "things that make life worth living" and includes such items as Groucho Marx, Willie Mays, the second movement of the Jupiter Symphony, Cezanne's apples and pears, the crabs at Sam Wo's, and Tracy's face, perhaps he is making the same point. Our enjoyment of life's mental riches, even if they are nothing more than figments, inexplicable ephemera flickering on the mind's palette in ways that we cannot even localize or identify, let alone explain, might be all that is required to justify, in the Darwinian sense, why the "hard problem" of consciousness arose in the first place.

But just as our superdeveloped sense of self-awareness brings us untold mental riches and can turn our most pedestrian lived moments into kaleidoscopic experiences of dazzling sensations and richly felt emotions, it carries within it the seed of darkness. Almost every joyful moment of self-awareness is undercut by the sure knowledge that these moments will not last forever. The cost of our peculiar means of existence whereby we can stand on both sides of the fence dividing subject and object is that we know we will die someday. In his incisive analysis of the human condition, *The Denial of Death*, Ernest Becker has argued that this deeply felt and uniquely human fact of our lives is the real key to understanding human nature.[12] Indeed, it isn't difficult to read important elements of Western creation mythology, such as the story of Adam and Eve's expulsion from the Garden of Eden, as a straightforward description of the consequences of our discovery of our own mortality. When we partook of the fatal fruit, the consequence was not just that we opened the window of self-awareness, but also that we understood that it would end. This is no small thing. Nicholas Humphrey, again writing in his book *Soul Dust*, suggests that the psychological impact of death awareness is so profound that it might even have been difficult for human beings to carry on at all. Knowing as we do that the curtains will come down at some point in the future, how do we muster the will to hoist ourselves out of bed every morning? Humphrey even goes so far as to suggest that several mysterious die-offs in the population of early humans that have otherwise been difficult to explain may have resulted from the inability of beings with a newly emerging sense of self to have been able to cope with this inescapable fact.

Given the pall of death that threatens to overcome our revelry in the panorama of our inner lives, it is perfectly reasonable to suppose that we may have developed explicit strategies to deal with death awareness. It's even quite possible that these strategies have been instantiated in specialized brain circuitry designed to lead us away from the sharp edge of despair and back to a productive life—productive in the Darwinian sense that we can shrug off the malaise of mortality long enough to look after ourselves, find mates, and reproduce. The simple fact that I am here writing these words and you are here to read them suggests some measure of success in this.

There are really three types of coping strategies, all outlined by Humphrey and all strategies that we can see in wide use in everyday life. First and foremost, of course, is simple denial. Although we all have had our share of 3:00 A.M. moments of despair in which we might sense the uselessness of the struggle in the face of the coming end, these kinds of thoughts are, for most of us and quite happily so, not those that motivate (or demotivate) everyday existence. When we are enjoying the sensations of a good meal, a wonderful holiday, or the feeling of warmth from the smile of a loved one, thoughts of mortality shrink into the background. For all intents and purposes we feel and act as though there will *always* be a tomorrow.

A different kind of denial, and one in which the majority of living humans participate, is the belief that our existence will, in some form, survive the death of our bodies. In other words, most of us believe in some kind of an afterlife. The form of this belief can vary dramatically from a conviction that we will be reincarnated to continue some kind of earthly voyage in new form, to the Christian belief that our sojourn on Earth will be followed by a very different kind of adventure—one in which we will surrender our bodies, but somehow hold on to our selves in a new and immortal form.

The third kind of response to our awareness of our own mortality is of a different type entirely because it constitutes less a form of denial and more a recasting of our understanding of our own relationship to the universe. This response involves a kind of mental gymnastics in which we convince ourselves that we are a part of something greater than that contained by the confines of our body. In this case, we identify with a culture or a set of institutions that both existed before our particular bodies were on the scene and will continue to exist after we die. It is this third kind of response to mortality that speaks most loudly to questions pertaining to our relationship with our built environment. In fewer words, we cope with the knowledge of our death by building, quite literally it seems, a legacy that will survive our departure.

There's a certain kind of logic to the list of possible responses to mortality that I've described. Each of these reactions to our certain knowledge that we will die makes intuitive sense and it's likely that most readers will even recall circumstances in which they have resorted to one or more of these patterns of thought in response to our anxieties about our own inevitable demise. But what does the evidence have to say? How do we even go about conducting psychological experiments related to our defense reactions to thoughts of mortality? One prominent research group, led by psychologist Sheldon Solomon of Skidmore College, has developed a comprehensive theory that is meant to explain how our death fears influence our everyday behavior, and especially how even unconscious impulses related to mortality might change our attitudes, beliefs, and prejudices.[13] The theory itself, called terror management theory, is solidly grounded in Becker's earlier argument that our knowledge that we will die propels culture. But the theory goes further by describing a mental state referred to as mortality salience, which is a state of mind that arises when something in one's environment produces intimations of mortality. In a typical mortality salience experiment, participants are primed to think about death by any one of a variety of means (writing a description of what happens to the body in death, reading a list of death-related words like *coffin* and *funeral,* or even simply walking past a funeral home). Following exposure to a mortality salience condition, the behavior of the participants is probed. The behavior probes used by this group are marked by ingenuity and panache. In one experiment, participants were told that they were required to complete a problem-solving task. The problems involved either hammering a nail into a wall to hang a crucifix or sifting some black residue out of a powder. The participants were provided with some tools that they could use to solve the problem, but in the most crucial conditions, the only tool that could be used to hammer in the nail was the crucifix itself and the only material that could be used to filter the powder was an American flag. The experimenters found that those people who had been primed with mortality salience cues were more reluctant, compared to a control group, to use a crucifix as a hammer or a U.S. flag as a sieve.[14] Although this finding (indeed perhaps the entire experiment) might seem a little bizarre, the authors argue that the greater reluctance to use sacred or revered items in a manner that might seem sacrilegious or even blasphemous suggests that the mortality salience condition had changed the world-view of the participants in such a way as to be more reverent and conservative. More generally, the pattern of findings in hundreds of experiments suggests that when mortality salience primes are used, participants generally respond by becoming more conservative and more attached to their own cultures and less tolerant of others. Indeed, terror management theorists have argued that it was the mortality salience prime induced by the tragic events of 9/11 in the United States that prompted not only the greatest boom in the sale of flags in the country's history, but probably even the re-election of George W. Bush's ultraconservative Republican government.[15] Collectively, these results suggest that events that bring our terror of our own death closer to the surface can exert sweeping effects on behavior, which could include not only changes in how we feel about other people, but perhaps also some of the incredibly expensive and time-consuming building projects that have resulted in the Angkor Wats, Great Pyramids, and Chartres cathedrals of the world. Whenever our existence is threatened and we are brought closer to the terror of our impending and inevitable death, we respond by shoring up our relationship with enduring culture, including its physical artifacts, resting on our confidence that though our fragile bodies may only last for the blink of an eye, the cultures to which we belong and contribute will last for considerably longer.

The arguments of terror management theory and the empirical demonstrations of the powerful effect of mortality salience manipulations on human attitudes and behavior, along with the threads that connect the appearance of might with evolutionary success, can help us to understand how we come to build anything bigger than a humble abode to sustain life and perhaps a collection of small shops and markets for commerce, but we still need to take a closer look at exactly what kinds of things happen when we step into a massive space like a cathedral or even an impressive bank headquarters or a large courthouse. The very existence of those buildings may be seen as a response to our terror of death, but what, precisely, happens to an individual who steps into one of them? What is awe for? Until quite recently, psychologists have been mute on such questions, but several recent studies lend some clues.

In an experiment that captured a great deal of public attention, Stanford psychologist Melanie Rudd exposed participants to awe-inducing experiences consisting of a short video presentation of people encountering majestic natural sights such as waterfalls, whales, and scenes of space exploration. In a control condition, participants saw scenes of happy parades and confetti trickling through the sky. Based on previous observations that awe experiences can affect our perception of time and produce a feeling of living-in-the-moment, Rudd designed a series of survey questions focusing on the subjective experience of time. Rudd showed convincingly that the experience of awe induces a kind of subjective time dilation. We feel as though there is more time to get things done, that the subjective instant is slowed. As a follow-up consequence of this perhaps, those who experienced awe were also more willing to engage in certain prosocial behaviors that, in her experiment, were contextualized in the form of expressions of willingness to donate money to worthy causes.[16] Although these findings do not fit easily into the framework I described earlier in which awe is said to produce an effort to accommodate, to bring into alignment mutually contradictory ideas about one's existence, it is possible to see a connection between Rudd's findings of time dilation and my earlier description of such phenomena as the Overview Effect in which we feel a sense of expanding space and a dissolution of the divide between ourselves and the rest of the universe. Just as the spatial divides of the universe might break down during the experience of awe, so one might expect to see shifts in the temporal horizons of our lives.

In even more recent experiments, psychologists Piercarlo Valdesolo and Jesse Graham demonstrated that the same kind of exposure to awe-inducing movies also heightened beliefs in supernatural agents and a lack of tolerance for randomness. In other words, after being presented with displays that produced awe, participants expressed stronger religious beliefs in an invisible and omnipotent hand at play in the design of the universe, and an unwillingness to accept the idea that majestic awe-inducing natural constructions could have been produced by random processes rather than by the guiding hand of something like a god.[17]

Although neither of these experiments involved presentation of participants with the sights of awe-inducing buildings, it seems likely that a wander through a massive cathedral could produce some of the same kinds of shifts in attitude and feeling, and it isn't difficult to accommodate such findings to the argument I put forward earlier that the purpose of such buildings may be both to encourage us to behave in a manner that promotes the success of the group (the increase in prosocial behavior) and to assuage our fear of mortality by increasing our willingness to believe in an omnipotent overseer who may offer us the promise of an afterlife.

The construction of large edifices and the experience of awe-inspiring architecture relates strongly to certain elements of human nature concerned with the maintenance of power relationships and social order. Some of these relationships can be seen easily enough as being evolutionarily continuous with basic processes in animal behavior that help to sort out aggressive instincts and dominion over territory. The big bank building that sprawls across an entire city block may stand in proxy for the vicious attack dog that may never need to do more than bare its fangs to assert dominance. On the positive side of the equation, the presence of a massive structure that represents the might of our money system or our judicial system may help us to feel secure. As we stand beneath the massive columns of a courthouse foyer or a temple, we may feel a tickle of fear to be in the presence of something greater than ourselves, and with the potential to overcome us as easily as we can swat a fly, but we might also feel the same kind of security that a young child feels when standing beneath the sturdy legs of a protective parent. To the extent that the structures that rise above us are believed to be benevolent, we feel less exposed to risk. The neural processes that underlie these kinds of basic animal responses probably reside in ancient subcortical brain circuits that mediate fear. Although the connections between our deep emotional needs and the vastness that encompasses our body may sometimes be more symbolic than real, the chains of cause and effect that make us feel subdued, conservative, and conformist in the presence of greatness are little different from those that might make a young or weak animal toe the line in a herd led by a more imposing alpha male.

Beyond this, there is something special about the human makeup. We know we are ourselves and in every moment of our lives except while asleep we are exquisitely aware of a subject–object divide in which we are both a part of, but entirely separate from, the rest of the universe. Even if we have complete faith that all of the facts of our existence, including our remarkable self-consciousness, are brought about by physical forces and the movement of atoms, we are deeply aware that our inside world feels as though it is made of different "stuff" than everything else. And in turn, out of that awareness of ourselves, springs our terror of death, which prompts us to turn to the supernatural, to cling to culture, or sometimes to simply deny what at a deep level we know to be inevitable. We turn to great buildings in part to dampen our terror, to experience awe with its space- and time-dissolving magic. This kind of mental alchemy, in which we use vastness to dissolve the boundaries of space, must also be undergirded by neural architecture, but it is unlikely to reside in the lower reaches of the nervous system. To understand how such awe responses are produced at the level of nerve cells and synapses, it seems obvious that the place to start is with a consideration of our awareness of our own body—its size and shape and the margins that separate our corporeal selves from the rest of the universe. For, surely, if we can identify any mechanism whose function might be responsible for phenomena such as the experience of the Overview Effect, the mystical union of one's individual self with infinite space and time as is represented by the crumpling bodies of pilgrims visiting massive temples and cathedrals, the answer is likely to begin with an understanding of how we know our own bodies in the first place.

Begin with something very simple. If I ask you to close your eyes and raise your arm over your head, you don't need to *see* your arm in its new position to know where it is. To start with, a rich network of receptors in the joints and muscles of your arm respond to the movement to help your brain keep track of your position. Not only this, but your brain also stores a copy of the movement command that you used to hoist your arm in the first place. All of this information coheres nicely to provide for you a good, accurate, and quickly updated representation of the position of your body in space. The beginning of the hard work of updating your neural representation of your body position takes place in the arm itself, is elaborated by neural circuits in the brainstem, and finally culminates in a richly realized version of your body in your cerebral cortex. The evidence for this comes not only from imaging experiments that can directly reveal the locations of such body representations, but also from the outcomes of certain types of brain damage that result in bizarre pathologies in which, for example, a patient might no longer feel that a body part actually belongs to her. In one disorder, known as somatophrenia, and one of a family of body-awareness disorders, patients will deny ownership of a part of their own bodies, claiming that the affected limb has been lost or stolen and that in its place is the limb of a relative, a member of the hospital staff, or even a different kind of animate object such as a snake. Large lesions that affect areas in temporal and parietal cortex, usually involving areas of frontal cortex as well, produce this disorder.[18] If nothing else, such cases demonstrate that we have networks of cerebral cortex whose job it is to maintain in good order our understanding of the scope, limits, and ownership of our own bodies. Put bluntly, we know which bits of the world, including the assortment of arms and legs that might be near us at any time, belong to us and which ones don't, and we have brain networks that normally do a nice job of sorting out this important problem.

In a way, this might seem to be a simple problem to work out. We can feel our own bodies, but not others. We've spent our lives in perfect alignment with the capabilities, extent, and reach of our own bodies. We are intimate with ourselves. But many other kinds of studies have suggested that this mental mapping of the space of our bodies is remarkably plastic. In one powerful demonstration, called the rubber hand illusion, participants are shown a rubber model of a hand. They see the hand being stroked gently by the experimenter and, at the same time, they feel their own hand being stroked. In a short period, participants come to inhabit the rubber hand as if it were a part of their own bodies. If, for example, the experimenter unexpectedly raises a hammer to strike the rubber hand, participants will flinch and show physiological fear reactions as if it were their own hand that was about to be pummeled.[19]

In my laboratory, we have used the tools of virtual reality to explore the limits of the embodiment of novel body parts. In one study, we had participants wear a helmet that showed them an ordinary indoor scene in which they could see a digitized version of their own arm, but with an important catch. We made their arm extend far from their bodies, with an ordinary looking hand extending from a long, telescoping forearm. As participants moved their own arm, they saw corresponding movements of the superlong virtual arm. Similar to the rubber hand illusion, participants quickly came to embody their crazy new arm as if it were a part of their own body and they showed appropriate perceptual and emotional responses to its use (in our study, rather than using a hammer, we had a very large virtual hypodermic needle approach the virtual arm as if to administer a giant injection. Participants didn't like this . . .).

In even more impressive studies, Henrik Ehrsson of the Karolinska Institutet in Stockholm and Olaf Blanke of the Brain–Mind Institute in Lausanne, Switzerland, have been able to produce remarkable out-of-body illusions in participants by allowing them to see images of their own bodies from remote locations (by using remote cameras in conjunction with a virtual reality helmet, participants looked on at their own bodies from a distance of a few feet).[20] After a short period, many participants began to feel as though they had left their bodies entirely and were looking on from another location. More formal tests of their perceptions of the location of their own bodies were consistent with these feelings. Imaging studies have shown that the brain areas involved in this illusion have considerable overlap with the areas affected in body awareness disorders like somatophrenia.[21]

Collectively, these kinds of studies suggest a surprising result: the mental representation of the margins of our bodies that we carry around with us from early in life and that we use to manage all kinds of interactions with our environment are extremely plastic. We can not only manipulate our perceptions of the size and shape of our bodies with a simple induction procedure that lasts for a few minutes, but there are many signs that we feel a genuine sense of ownership of these newly shaped bodies. In many other cases of brain plasticity, it isn't hard for us to understand why it makes sense for our nervous system to be malleable to experience. For example, the coordinated processing of information from our two eyes, called stereopsis, is responsible for much of our ability to see in depth. Stereopsis develops fairly early in life, but it depends on experience for proper calibration. Given that this calibration depends on the size of our adult heads and the exact distance between our eyes, it makes sense that experience should play a role. But what about body awareness? Certainly, we would expect to see some plasticity in our body representation early on in life as we grow, but as fully formed adults, this strange kind of ability to instantly retool our conception of the form and location of our body makes less sense. Why should our self-awareness be able to ooze into body parts of new sizes and shapes with such breezy alacrity?

According to Fred Previc, author of *The Dopaminergic Mind in Human Evolution and History,* an answer to this question might reasonably start with a starkly simple observation: when we consider the far reaches of distant time or space, our gaze turns upward.[22] Asked to do a complicated mathematical task in our heads, we look upward. Our eyes roll upward when we think of large spaces or distant times. Interestingly, upward rolls of the eyes also frequently accompany intense religious experiences, meditative trances, and hallucinatory states. Finally, of course, when we enter a building of vast scale, our eyes turn toward the ceiling. Indeed, in some kinds of religious architecture, design elements are deliberately crafted to encourage us to peer upward toward the apex of a building that appears to stretch to the height of the heavens. In Gothic designs, for example, repeating elements at a range of scales (much like the fractal designs described earlier) are used to help generate illusions of immense height, as if the top of the building extends to heaven itself.

To understand how upward gaze relates to an intimate connection with the infinite, Previc says, we need to consider the overall organization of brain systems that monitor space. In common with all other mammals, human beings possess a collection of brain systems that are charged with the surveillance and control of peripersonal space—that is, the space that immediately surrounds the body and is within easy grasping reach. Not surprisingly, this part of space lies mostly below the horizon line of gaze. The space that lies higher in the visual field—above the horizon—is typically the part of the world that contains objects that are well outside grasping distance. This zone of so-called extrapersonal space extends far off into the distance and it contains important information that we might use to plan elements of our behavior that go beyond the immediate confines of body space. According to Previc, it is no accident that in humans the neural representations of extrapersonal space are much more prominent than they are in any other animal. These same systems are responsible not only for coding our far-off gazes of distant places, but they also house some of the machinery that is responsible for abstract thought and reasoning. Conceptually, this makes some sense as abstractions, by definition, involve mental processing of information that lies beyond the here-and-now. Hardware that is specialized for transcending the intimate bounds of the body and grasping space would be ideally suited to such work.

Thinking of the overall scheme of the brain's organization in the analysis, comprehension, and control of space in this way helps to make sense of a wide range of human psychological phenomena, many of them related to our feelings of transcendence and spirituality. For example, excessive activity of dopaminergic brain systems, such as those seen in schizophrenia, can result in distortions of our normal understanding of our relationship with the world. Hallucinations and delusions, many of which prompt us to believe that the normal physical trappings of our relationship with the world no longer apply, can be considered as abnormalities of extrapersonal systems normally charged with working out the relationship between ourselves and the greater world. Intense meditative states, in which practitioners report a dissolving of the boundaries of the body and a feeling of oneness with the universe, have also been shown to correlate with a tipping of the balance of neural activity away from peripersonal systems and toward these special dopaminergic extrapersonal systems. Lesions that affect areas of the brain that participate in peripersonal systems also shift the balance of activity toward extrapersonal information processing, and some studies have shown that those who suffer such lesions also experience intense religious and transcendent experiences.[23]

When we walk through a grand space, be it a stunning natural vista, a large cathedral, an impressive city hall, or a courthouse, one of our near-universal responses is to gaze upward. As surely as a mystic may focus upward to engage a "third eye" or an individual engaging in intense prayer may try to focus their attention on the heavens above, this upward focus of attention activates an extrapersonal information processing system that primes us to focus on the faraway, the distant, or even the infinite. This upward focus helps us to dissolve the earthly chains that bind us to the prosaic events of ordinary life, the exigencies of mere survival, bodily sustenance and protection, and ultimately our awareness of our mortality, and to feel the positive emotion and comfort that come from connection with a greater existence—some would feel even a divine one. Though there are many other complex feelings and adjustments of behavior that might be generated by the experience of being enveloped by vastness in the built environment, some of them enjoying evolutionary continuity with the effects felt by other animals as they find their place in a social order or feel the protection of a powerful parent, the operation of brain systems that encourage us to feel contact with the sublime and celebrate the miracle of self-awareness are supremely and uniquely human. It may be here that we find the secret formula that allows us to poise miraculously on a knife-edge of existence, enjoying all of the benefits of the inner theater of our minds while at the same time coping with the abyss of our own eventual certain deaths. Perhaps it is here that we find the most dazzling exhibition of the power of the built environment to sustain our fragile purchase on such a narrow ledge.

CHAPTER 7

SPACE AND TECHNOLOGY: THE WORLD IN THE MACHINE

O N A COOL AUTUMN DAY IN 2007, I arrived in the beautiful town of Santa Barbara, home of one of the University of California's stunning oceanside campuses and a cluster of high-tech industries, mostly spinoffs from campus research. I was there to visit a group of young entrepreneurs who, under the leadership of a smart psychologist named Andy Beall, were in the beginning stages of building what has become one of the world's leading purveyors of virtual reality hardware and software systems. After a brief pitch to my university dean, I had been stunned when, over the course of a couple of days of phone calls, he had found me a substantial sum of money to build a VR lab. I was so frightened of buying the wrong system that I decided to use some of my own money to visit the company to see whether their products lived up to the marketing hype. I sat with Beall and his colleagues in a generic warehouse space located off a dingy back alley near the city center, half-listening to his description of the company's origins and mission. As friendly and interesting as he was, I hadn't come all this way for speeches—I wanted to play with the toys.

Eventually, the group led me through a labyrinthine series of hallways into an empty room containing some large headsets (called head-mounted displays or HMDs), a few computers, and little else. Soon, I was wearing an HMD and over the course of about an hour, I found myself teetering on a narrow plank suspended over a deep pit, racing toward a precipitous cliff in a fast, red sports car, standing on the platform of a train station as a high-speed train surged toward me through a tunnel, and tracking an armed soldier through a scorched village some-where in the Middle East while he tried to shoot me.

As I write these words, it occurs to me how much the world of com-puting and visualization has changed over a few short years. What, at that time, was an experience that I was able to dine out on for weeks, now sounds as if I'm describing the typical experience of a teenager sitting before a gaming console on a rainy weekend afternoon. Yet, for the moment, but probably only for a short while longer, there are some differences between the capabilities of a research-caliber virtual reality system and even the most advanced gaming systems. Virtual reality sys-tems that use HMDs with advanced graphics capabilities and comput-ing are able to provide a panoramic, 3D simulacrum of the real world that users find so compelling that they become fully engaged with what they are seeing. The experience of immersion, in which the senses are filled completely with the sights and sounds of the simulation, gives rise to the holy grail of virtual reality: the feeling of presence. Presence is exactly what it sounds like: we lose touch with the "real" world outside of the helmet and we feel embodied in the simulation. Simulations of risky environments such as Virtual Iraq, a scenario that provides the visitor with the sights and sounds that a soldier in the Middle East would experience during the heat of battle, are sufficiently realistic that the heart pounds, the palms sweat copiously, and the body courses with adrenaline. The simulation is so compelling that it has been used with some success in the treatment of post-traumatic stress disorder for sol-diers returning from warfare.[1]

It isn't easy to convey the experience of high-fidelity virtual real-ity, other than to say that it's a bit like lucid dreaming. Participants never completely lose awareness that what they see in front of them is a kind of perceptual trick. They can hear the real-world scraping of shoes and the murmurs of the technicians who work the controls, just as one might slowly become aware of the sounds as one awakens from a dream. But the images and sounds that are presented on the screens of the HMD can elicit visceral emotions—fear, surprise, and exhilara-tion—and postural defense reactions. Placed in the driver's seat of a car, it isn't easy at all to overcome the psychological resistance that prevents users from plunging off the edge of a cliff even though they know that it is imaginary and that nothing will happen if they do. Walking around in tight spaces, one naturally leans away from walls and ducks under low-hanging obstacles, even in the full awareness that everything that is being seen is composed of nothing more than pixels.

In a way, these simple descriptions of the phenomenology of the virtual experience resonate with some of my discussion in the last chapter. Something about the makeup of our nervous system makes it remarkably easy for us to jettison the trappings of real life and to take flight into journeys of the imagination. Just as we can be convinced to morph our representations of our bodies into weird shapes, or even to leave our bodies entirely, we can also be invited easily into the imaginary world of the virtual.

On the same trip to Santa Barbara that saw me fighting virtual Iraqi soldiers, I had a chance to visit with University of California social psychologist Jim Blascovitch, one of the first scientists to recognize the staggering potential of virtual reality for psychological experimentation and still one of the leaders in the field. Sitting in his office in front of a huge picture window that framed a magnificent view of the Pacific Ocean, it was a little difficult at first to even focus on his words. It was ironic that Blascovitch was describing to me the prepotent ability of the human mind to wander. Diary studies of mind-wandering conducted using specially programmed smartphones that prompt users at random times to indicate their mental state suggest that we spend as much as 50 percent of our time disengaged from our current activity, and that our mind may wander to another time or place, on average, about once per minute.[2] In my earlier book, *You Are Here*, I argued that this propensity to live outside of the here-and-now, though it might cause our intimate relationships with certain kinds of spaces to become strained, has facilitated the uniquely human ability to envision, project, plan, and build—in other words, the wandering mind may be a key part of the human cognitive makeup that has enabled material culture.[3] In Blascovitch's view, these irresistible mental twitches are also related to our proclivity to become engaged in virtual environments. Mentally, we live in a state where we are always halfway out of our chair and on the way to some other place, so putting us into a compelling 3D simulation can almost instantly transport us to a feeling of presence in a new synthetic environment. For humans, Blascovitch pronounced as he sat across the table from me with a Cheshire cat grin, "Everything is virtual."

Experiments by Blascovitch's collaborator Jeremy Bailenson of Stanford University suggest that even environments that are only slightly immersive can produce behavior that suggests some degree of presence. In the metaverse (or metaphorical universe) called *Second Life*, a giant, open sandbox in which people can embody avatars—simple humanoid forms that can be embellished with personalized facial features, clothing, and even novel anatomical appendages—visitors can walk from place to place, explore virtual buildings, and most importantly, interact with other avatars, each of which is embodied by another visitor to the metaverse. The designers of *Second Life* have included some simple tools whereby it is possible to learn a few things about the other avatars that one encounters: their names, genders, and short biographical sketches. It's also possible to measure the characteristics of social interactions, including the orientation of the body and the distance between two conversing avatars. In one creative experiment, Bailenson and his colleague Nick Yee "stalked" other avatars and watched their interactions. Yee and Bailenson discovered that the body language of conversing avatars followed the same rules as social interactions in real life.[4] For example, rules of proxemics worked out by sociologist Edward Hall and described in his book *The Hidden Dimension* showed that two men in conversation tend to stand further apart than two women or a mixed-gender dyad. During conversation, men are also less likely to make full eye contact than are two women or a man and a woman.[5] In *Second Life*, these same rules of proxemics apply to conversations between avatars. Conversing men maintain greater pixel distances between them than conversing women, and their bodies arc pointed at oblique angles to one another rather than in direct frontal contact. When one considers that the avatars are low in graphic sophistication, looking much like cartoon characters, and that the operator of an avatar doesn't employ a true first-person perspective (while moving through *Second Life*, one normally has a vantage point that is slightly above and to the rear of one's own virtual body, very much like that experienced in Olaf Blanke's out-of-body simulations), this is further strong evidence for the ease with which the location of our physical identity can slip around in space, easily taking up residence on other places, even ones made only of computer code.

Some of the most creative uses of immersive virtual reality in research have arisen in social psychology laboratories like those of Jim Blascovitch and Jeremy Bailenson, in part because of the ease with which people seem to become transported to new environments while bringing some elements of their personal identity along with them.

In one experiment, led by Mel Slater of University College in London, participants who were fans of the Arsenal football club in England were brought into the laboratory and placed in a virtual simulation of a British pub. While there, they met an avatar who struck up a friendly conversation about football. There were two conditions. In one condition, the avatar professed to be a fan of Arsenal, while in the other control condition the avatar was a fan of some other team or sport. At some point in the conversation, a third party—again an avatar—entered the scene and began to pick a fight with the original avatar. The dependent measure was the extent to which the participant—the only real person in the study—attempted to intervene in the dispute. When paired with an Arsenal avatar, the video records show the human participant becoming increasingly disturbed by the altercation, sometimes even trying to throw his very real body between the two simulated combatants. When paired with a neutral avatar, there was much less intervention by the human observer.[6] This remarkable demonstration shows the power of an immersive simulation to capture the attention, affect, and reflexive social conditioning of a human observer over a remarkably short period. Such environments represent a tremendous opportunity for researchers in human interaction. Where the history of this discipline has mostly involved the use of human confederates whose behavior might reasonably be expected to vary a bit from trial to trial, a simulated person can be controlled in every detail of behavior, and can be relied upon to do exactly the same thing in trial after trial.

In a somewhat more applied realm, Blascovitch has explored the role of spatial location in the classroom on learning processes in students. Not surprisingly, real-world research has shown that students learn better and retain more information in a classroom when they are seated in the center of the room and near the front. From there, they have close contact with the instructor, frequent direct eye contact, and generally closer engagement with the ongoing lesson. Blascovitch realized that with the use of virtual environments, space could be bent so that every student in the class could sit in the same location—the front and center sweet spot. When he developed a virtual environment that produced this effect, students taking up the prime position in a virtual classroom showed the same learning benefits that have been shown for real bricks-and-mortar settings.[7]

In my own laboratory, where our focus has been more on the influence of built settings on behavior, we have been able to take advantage of immersive virtual environments to learn more about the human response to variations in the geometry and surface textures of built settings at every scale from the insides of buildings to vast urban streetscapes. In experiments in our lab, we have used such methods to explore the responses of participants to compelling simulations of existing homes. Those experiments showed a remarkable correspondence between our predictions about how people would move through and express preferences for locations within a home based on its geometry and the resultant patterns of movements of participants in the VR model. As a test bed for our own theories related to the psychology of architecture, the ability to be able to create an entire full-scale house out of nothing but pixels is a fantastic boon. But what about in the real world of architectural design? Good architects of domestic spaces are keenly concerned with designing a home that fits the personalities and preferences of the future owner of the space. What the methods we have used might make possible is to put those personalities and preferences into a clear empirical context. Architects could render models of a design and essentially place their clients inside them to see how they respond and where they go. If they wished, it would even be possible to attach a few simple physiological instruments to the client to collect body-based data pertaining to their emotional state. In our own experiments, we used such measures to garner preliminary evidence that peoples' stated preferences for particular types of spaces were sometimes out of kilter with the readouts from their bodies and their movements. Such dissociations might make possible a fascinating, nuanced, and ultimately very useful portrait of a user's response to a space.

Using larger-scale environments, it is also possible to look at the responses of visitors to virtual cityscapes to measure their responses to places composed of many city blocks of space. In my laboratory, my student Kevin Barton has used such environments to explore the factors that influence wayfinding in cities. Barton built two kinds of large urban-scale environments. One of them consisted of a gridlike street plan with long avenues providing good visibility and straightforward intersections from which it was possible to look ahead and to plan easy routes from one place to another. Inside the environment, the visitor saw something a little bit like a Manhattan-style street plan, but with a few extra kinks in the streets to lend greater realism. The second environment contained more organic, serpentine streetscapes where visible routes from one place to another tended to be quite a bit shorter and many of the intersections were more complex. It looked a little more like a segment of the city of London, or perhaps New Orleans. In the language of space, we would say that the "grammar" of the two street plans was different. The grid plan would be described as being more "intelligible" where that word, though it still connotes the everyday sense of understandability, also has a precise mathematical meaning related to the manner in which streets are joined to one another and to the average complexity of a route (explicitly, the number of changes in direction) from any one location in the space to any other location. Although many experiments in wayfinding have been conducted in real streets and cities, the advantage of a virtual reality city is that it is possible to specify with exact precision how the streets relate to one another.

Participants in the experiment were required to find their way from an edge of the environment to an undisclosed location that contained a prominent landmark, which looked like a monument such as a cenotaph. Not surprisingly, participants found it much more difficult to locate the monument in the less-intelligible London type of environment than in the more straightforward grid setting. It took them longer to find the monument, and longer to find their way back to their starting point; there were also many more pauses and hesitations along the way. But with a virtual environment, we were able to go a little bit deeper into the minds of participants by using other careful measures of behavior. For one thing, we measured the frequency with which participants blinked their eyes, and whereabouts in the environment these blinks took place. It might seem a little strange to be interested in blinks, but other research has shown that we blink more frequently when we are engaging in difficult, effortful, cognitive processes. It's as if we are trying to help ourselves to think or turn inward to our thought processes by literally blocking out the external world. When we looked at blink rates in the virtual environments, the spatial distribution of blinks showed stark differences in the two environments. In the low intelligibility environment, we saw intense blinking episodes at many different locations whereas the blinks in the more intelligible environment were localized to one or two knotty locations. So using these methods we were able to produce a kind of visualization of the average amount of mental processing that was taking place in a participant for any particular location in the environment.[8]

There are other kinds of clever tricks that can be used in virtual reality to help to understand wayfinding processes. For example, another of our experimental questions had to do with how individuals make decisions about which way to turn when they arrive at an intersection when they're looking for some particular destination. How do they remember where they've been and work out where they want to go next? One idea is that they try to memorize the local features that they see at an intersection. In the real world, those features might consist of prominent landmarks, signs, and objects. Our environments were devoid of such landmarks, but they did contain intersections that varied in their geometry (X-, T-, and Y-shaped intersections, for example). Another way that someone might try to solve a thorny wayfinding problem could be to look off into the distance—to peer down the streets that they can see from their current location to figure out what features were present at the next set of intersections that they could see. We were easily able to tease apart those two types of information in our virtual environments by introducing a little weather! We introduced a fog into the environments that made it impossible for participants to see beyond a certain distance, and we had precise control over exactly how thick the fog was and how far the participants could see. What we found was that the mix of local (the immediate intersection) and global (looking down available avenues) information that was useful to a participant varied depending on the overall design of the streetscape.

As with the experiments designed by Jim Blascovitch, Mel Slater, and Jeremy Bailenson, the work that we have done using virtual environments is both theoretically interesting (it tells us new things about how we deal with problems of space and generates new hypotheses) and practically useful. It isn't hard to imagine how a planner might be able to use the wayfinding visualizations that we've generated both to understand what is happening on streets as they are designed currently and also to glimpse into the window of future possibilities by tweaking designs to see their effect on minds. When I wear my researcher hat, I am excited by the possibilities engendered by new technologies that allow me to present participants with interesting and complex surroundings, but over which I have surgically precise control. Now it's time to remove my scientist hat, though, and to try to think like the average citizen. As I will describe, the technologies that we use in our laboratory are no longer restricted to the cloistered environments of a research lab. They are beginning to permeate our everyday lives, and this is a process that will accelerate rapidly over a few short years.

When I began to do research in virtual reality in earnest less than ten years ago, one of the factors that encouraged me to make the leap to a new kind of technology was an economic one. In times past, the cost of a research-quality virtual reality setup of almost any kind was stratospheric and only within the grasp of large groups of researchers or well-funded military laboratories. When I leapt into the fray, the cost of a good laboratory had dropped to about $100,000. This was still a lot of money, but it meant that the ability to transport participants into virtual worlds of almost photorealistic quality had dropped to a level that was affordable by many institutions. I only provide you with these boring facts of economics to make clear the seismic shift that is now beginning to take shape.

In 2011, a precocious eighteen-year-old Californian named Palmer Luckey, frustrated with the lack of availability of good-quality head-worn displays for virtual reality, put together an ingenious set of inexpensive components, along with a healthy amount of duct tape, to produce a prototype of a headset, called the Oculus Rift, that seems destined to outstrip the capabilities of many much more expensive headsets, including the $30,000 models that I use in my laboratory. After several design iterations, and a spectacularly successful Crowdstarter campaign, Luckey parlayed his initial successes into a product that is poised to see commercial release sometime in 2015. It will probably cost consumers about $300. As one indication of the momentous anticipation of the impact that this device will have on the world of virtual reality immersion, Luckey sold the company that produces the Rift to Facebook for an amount in excess of $2 billion.[9]

If the Oculus Rift lives up to the hype, and early users insist that it will, it will place into the hands of ordinary consumers the tools to immerse themselves in compelling immersive virtual environments on an everyday basis. While wearing a lightweight headpiece, we will be able to transport ourselves from our living rooms to essentially anywhere that can be photographed or modeled in full stereoscopic, high-resolution glory. It's a mug's game to try to predict the future, but the conclusion that the aptly named Rift will, quite literally, tear a big hole into the fabric of our lived time and space does not seem to be a gushing overstatement so much as a sober pronouncement on the future of display technology.

To understand the appeal of easily available virtual reality, begin by considering the massive surge in the popularity of computer gaming. Long gone are the days when games were the exclusive province of teenagers, mostly boys, who disappeared for hours at a time into fantasy worlds and first-person-shooters. According to the most recent statistics from the Entertainment Rating System Board, two-thirds of households in the United States have a gaming system. The average age of a gamer is thirty-four with fully 26 percent of the gaming market consisting of adults over the age of fifty. Women make up almost half of the market for computer games. Revenues for the gaming industry top $10 billion in the United States, while global revenue in 2013 was placed at over $90 billion.[10] By way of comparison, total revenue for the motion picture industry in the United States is approximately equal to that of the gaming industry, and this revenue share has actually been declining for the past several years. Compared with today's computer games with sophisticated graphics, strong narratives, and exciting engagement, the act of sitting passively to watch a story unfold on the big screen or on a home monitor or laptop may soon become passé.

Just as we crave the higher levels of engagement and action that are afforded by games, we also have a strong appetite for novel means of interfacing with games that go beyond keyboards or simple hand-held controllers. Devices such as the Wii and the Kinect, both of which allow a more natural gesture-based form of interaction with computer games, increase the range of different kinds of activities that are available to the gamer and increase our levels of immersion. Using these kinds of simple motion tracking devices, it becomes possible for us to make movements during gameplay that resemble the kinds of movements that we might use during the real-life analogs of the situations being depicted in the game.

In its formative years, the gaming industry suffered from a dire lack of engaging narrative. Though the graphics were pretty and the avatars surprisingly lifelike, the stories that they told were only impoverished cartoon versions of real life, and we most often only stayed in the game for the thrill of hair-raising shoot-'em-up experiences. But now this is changing as well, as game developers realize the importance of including story elements that are just as well conceived as the backgrounds and settings of the action. Although one might argue that, when it comes to immersion, even the best video games cannot come close to the thorough pleasure and escape that is offered by a great story alone—say a novel by Dickens or Tolstoy—the numbers alone suggest otherwise. People, especially the younger ones among us, are voting with their feet. Given the dramatic shift in audiences looking for entertainment and cultural enrichment through interactive gameplay, how long might it be before the best artists turn to such popular and exciting media? Though parents, educators, writers, and anyone else concerned with the preservation of traditional forms of expression may bemoan this development, there's no doubt that it is happening. Furthermore, given the potent demonstration of the allure of active game play when compared to more passive forms, there is every reason to believe that the trend toward interactive computer media for entertainment, education, and culture will accelerate when fully immersive, wide-field, 3D, motion-tracked virtual reality is available to all for less than the price of a good gaming console.

There's little question that the emergence of widely available VR technology will make possible a remarkable advance in the availability of compelling immersive experiences. Separated loved ones will be able to feel a compelling facsimile of face-to-face presence. With the steady advance of the specialized field of teledildonics, it's even possible that we will be able to engage in simulated touching, foreplay, or even a strange form of sexual intercourse using the instruments of virtual reality.[11] Students in classrooms will be able to put on a lightweight headset and find themselves transported to the streets of ancient Rome. Journalists will be able to go far beyond the simple presentation of words and images to convey stories. The meaning of "embedded journalism" will change entirely when the news consumer is able to essentially enter the heat of a war-torn site or the heartbreak of a humanitarian crisis.

Such things are already happening in a limited way: the University of Southern California's Interactive Media Lab has produced Project Syria, which allows viewers to experience the sights and sounds of a rocket attack in Aleppo.[12]

Given the prospects for radical changes in the way that we understand and move through space and all of the heady prospects for new kinds of experiences, it would almost seem curmudgeonly of me to raise the specter of a dark underside, yet the same technologies that will bring such advances in our means of travel, entertainment, and interaction will also bring risks, at least some of them related to the manner in which we relate to built spaces.

Metaphysical Disorientation Syndrome

Not long after I built my VR lab, I submitted my first ethics application for a proposed experiment to our university review board. A few days later, I received some curious feedback. One of the reviewers had asked a question: what if, following an experience in one of my beautiful virtual environments, a participant became so confused about the relationship between my experimental virtual world and the real world that difficulty distinguishing between the two ensued? What if participants became so disoriented that they were no longer able to carry out ordinary transactions in ordinary space? Would I agree to send them home in a taxi at my expense and check up on them later to ensure that they had safely reentered physical space? My gut response to the question was to emit a loud guffaw and to mutter to myself that, if my environments ever became so persuasive, I would pat myself on the back for a job well done! Yet, thinking back now on that experience, I can't help but wonder whether there might have been a little prescience in the question. Based on experiments using environments that in some cases, were not as meticulously detailed as the ones in use in my lab, we know that immersion in certain kinds of virtual experiences can produce enduring effects on the user. Those people shown younger, more attractive, or more physically fit versions of themselves in virtual reality are more likely to express the intention to look after themselves. Those given superpowers (such as the ability to fly) are more likely to engage in heroic, or at least more prosocial behaviors, following the experience.[13] In these cases, one could argue that, in the hands of the experimenters, the participants had actually become, in some way, *better* people. But if we can become better by virtue of our experiences in virtual reality, then surely we can also become worse. What is the long-term effect of exposure to violent acts in an immersive environment? How does being embedded in a rocket attack in Syria change our identity? While the debate rages about the long-term impact of exposure to violent games of a less immersive flavor, I'm not sure that anyone has the answer to such questions.

More germane to a discussion of the impact of immersive virtual spaces on behavior, I think first of an experience of mine in an early demonstration of our laboratory to a class of architecture students. To impress them, I showed them a series of finely rendered models of full-scale home spaces that I had designed for an experiment. In the demonstration, the students were able to walk through the homes, explore their textures and viewpoints, and generally develop a sense of presence in the environments. Or so I had hoped. What the students ended up doing was climbing the stairs in one of the models and leaping out of the upstairs window to see what would happen. In this case, as my models weren't scripted for such offbeat actions, what the students managed to do was to crash my model so that they came face to face with the famous "blue screen of death" known the world over as a generally discouraging sign for a computer program. With some hindsight, I should have expected no less from these creative and adventurous students, immersed as they were in all kinds of methods for high-performance visualization of 3D environments. But the deeper question, as we see increasing penetration of synthetic environments in our lives, is whether we will come to see the normal constraints and barriers afforded by physical space to be less important. If the experiments showing carryover of effects from virtual reality into the real world are true, then this seems likely. I'm not suggesting that we will all begin leaping out of windows or off building roofs with alacrity, but more that our understanding of the manner in which physical space normally constrains movement and activity will begin to shift in directions that may not be easy to anticipate.

In my previous book, *You Are Here*, I wrote about the potential for any kind of space-shifting technology from telephones to television or even high-speed travel in airplanes or trains to weaken our already fragile grasp on geographic space, the connections between things, and the order of physical space. Our mental readiness to be catapulted from one space and time to another conspires, in the case of immersive virtual reality, with a nervous system that is designed beautifully to adapt to changing patterns of stimulation. For one signal indication of the extent to which the extraordinary plasticity of our brains may make us vulnerable to chronic exposure effects in virtual reality, we need only reach back in time to one of perceptual psychology's classic early experiments. In studies conducted at the end of the nineteenth century, Berkeley psychologist George Stratton wore a special set of prism glasses that produced left–right and up–down reversals of the visual images arriving at his eyes. One might think that such a bizarre form of visual input would produce severe disorientation and difficulties in moving about in the world; indeed, this is exactly what Stratton discovered. But, remarkably, over the course of a few days, Stratton adapted to the glasses completely and found he could use vision as he normally did. Upon removal of the glasses, Stratton once again found himself disoriented and required a short period of re-adaptation to the normal pattern of visual input to his eyes. If our brains possess sufficient plasticity to learn to adapt to such severely disordered visual information after just a few days, it doesn't seem unreasonable to suppose that we might be prone to similar kinds of space-bending illusions after repeated experiences in virtual reality.[14]

In a similar vein, William Warren and his students at Brown University in Rhode Island built a set of peculiar "wormhole" environments in immersive virtual reality. These environments included several portals where, as participants walked through them, they were instantly transported to a new location in the environment. Anyone who has played the computer game Portal will instantly recognize the effect. Surprisingly, participants were able to construct mental maps of the environment that made a good deal of sense even though they were completely unaware of the existence of these strange warps in the geometry of the spaces they traveled through. For Warren, the conclusions of these experiments had to do with the manner in which we mentally represent spaces. We are much more in tune with the connections between spaces, the topology of space, than we are with old-fashioned Euclidean geometry.[15] But for a world poised to become increasingly penetrated by technology for generating immersive environments, findings such as Warren's again suggest the possibility that we may become more muddled than we already are about how space works and the influence that it plays in our lives.

WYSINWYG (What You See Is Not What You Get)

One kind of relaxed response to the future possibility that we may find ourselves immersed in a peculiar amalgam of real and virtual spaces as we go through our daily lives is to suggest that there's nothing here to fear and everything to look forward to. Who wouldn't want to avoid a long, boring journey by popping through a magic rabbit hole to a new destination? Isn't that much like the allure of the fantastic transporter beams used in *Star Trek*? And to continue with the same metaphor, who wouldn't want to have easy access to something a little bit like the holodeck? Wouldn't such a thing be both a fantastic source of entertainment, but also a useful resource for training, education, business, travel, and even intimate interpersonal interactions?

A more measured response, though, would have to include a consideration of the possibilities for the bending of space that may be less benign. We've already seen some examples of this. Casinos are built to empty our pockets. Designers of shopping malls and department stores pay particular attention to the manner in which environmental variables can promote impulse purchasing. Environments that encourage us to think about mortality may cause us to embrace social conservatism. With physical buildings, the designer is limited to a set of measures that are relatively resource intensive (it takes time and money to build walls and ceilings) and that must largely cater to the average person (everyone sees the same physical structure). Virtual environments lift both of these constraints. Virtual environments can be constructed and modified in an instant because they are only made of the photons that are emitted by a display device. Further, because each viewer is immersed inside his or her own version of an environment generated by a personal device, each person can be presented with an environment that is completely adapted to that viewer's history, tastes, and interests. Earlier, I mentioned that Facebook had purchased the Oculus Rift company for a price in the billions of dollars. Though we might like to think of Facebook as an application that is designed for us to keep in contact with our friends, the real raison d'etre of the application is to encourage us to buy things by means of personalized marketing. Instead of an irritating advertisement that appears on the screen beside the check-ins of your friends, imagine if the products that were being promoted appeared in 3D splendor, forcing you to almost literally step over them to advance to your real goal. Further, imagine that you are looking for one of your friends in a 3D environment and you get lost along the way, suddenly realizing that your newsfeed has become something a bit like the confusing aisles of an Ikea store, except that every item that you pass in the store is somehow related to your recent interests, as measured using online snooping of one kind or another.

Regardless of the particulars of what Facebook intends to do with the Oculus Rift partnership, it is hard to believe that for the amount of money that has changed hands in the transaction, there is not a strong intention to find a way to merge virtual environments with a social networking site in order to make money by tapping into our preferences and our passions.

First-Person Shot

The emergence of self-consciousness is a key development in some of our uniquely human responses to built environments. The very essence of that self-consciousness is the privileged, insider view of the world that we all possess and know as the first-person perspective. In literature, film, and even in computer games, we make distinctions between this unique perspective and other perspectives that are not centered directly on the observer. There may be individual zones of influence for this intimate perspective, such as Fred Previc's distinction between peripersonal and extrapersonal space, but all such zones emanate from the egocentric perspective. Although we may reflect on the first-person perspectives of others and what they might be like (and our being able to do this at all is one of the foundations of empathy), we are used to inhabiting a world where there is one clearly dominant first-person view, and that is our own. But think back first to Olaf Blanke's compelling demonstration of the out-of-body experience conjured with the relatively simple combination of a webcam and a virtual reality helmet. In one way, this seems a very simple shift of perspective—we feel as though we have shifted to a location a few feet behind our body, much as we might feel some sense of embodiment for an avatar in *Second Life* that we view from a short distance and control with a keyboard or a computer mouse. But there is another subtler sense in which, though we may have learned something important about the lability of our sense of embodiment, we have lost some of the preciousness of the first-person view. We've grown up with a certain sense of the dignity, privacy, and uniqueness of our own personal cockpit on the world; and then, fairly suddenly, and only through the intervention of technology, those qualities seem to have been sullied. If virtual reality technology becomes endemic as a user interface for visualization of the world, then our personal perspective can be placed anywhere at all—not only in inaccessible locations like Aleppo, but also *inside* strange things such as beings with ultralong arms or, as in experiments conducted by Jaron Lanier for Microsoft, into the body of a simulated lobster. Through the use of technology, our first-person perspective can actually invade the world and be anywhere that we want it to be. The implications of this are staggering.

We cannot consider being human as a kind of mind-in-a-vat experience—as if we are nothing more than a central processor like a computer whose connections to the world are arbitrary. Instead, the connections between mind and body—our postures, movements, our entire physicality—are not only essential to the development of feelings, but also of thoughts. This same physicality is also the key feature governing how we relate to everything in our environment, including not only built structures, but other human beings as well. What virtual reality technology has shown us is that the design of our minds, by jacking into our predispositions to take mental flight from one time and place to another, is such that the exact form of our embodiment can be shape-shifted. Whether by inverting goggles, immersive VR helmets, or the augmented reality of a device like Google glasses, our understanding of both the shapes of our bodies and where they begin and end are all open to modification. Although we are beginning to understand the science, we have not begun to consider the wider implications of these new discoveries for human existence.

I find it hard to achieve the right balance of feelings about these developments, both those already realized and those that lie on the horizon. On the one hand, I'm excited by the advent of technology that will make possible entirely new ways of understanding self-consciousness and embodiment. There is no question at all that these technologies are making possible revolutionary new experimental methods in psychology, of which I am currently taking full advantage. At a practical level, I can see the vast potential for new methods of conducting exchanges, building new relationships, visiting otherwise inaccessible places, and designing fantastically compelling environments for education. At the same time, I'm a little worried about the possibility that such technologies, as they take further hold, may engender a kind of cheapening of some of the things that I hold dearest about my life. If, as Nick Humphrey tells us, one of the main evolutionary benefits of self-consciousness is that the panorama of the sensory environment, experienced in a privileged and direct way, makes life worth living, then it feels as though the full emergence of technologies that will allow us to sling that perspective from one place to another, mold it to the requirements of any kind of stakeholder from a furniture store to a political party, involves a devaluation of the unique, individual experience. As I walk through a forest, peering upward at the dense foliage above me, watching the rippling effect of the wind, listening to the sounds of birds and of my own rustling footsteps on the ground, an important part of the joy of that experience is knowing that at that very moment in time, it is just for me. I've enjoyed a unique moment in time and space, one that can never be repeated. Perhaps in the same way that Walter Benjamin, in his essay "The Work of Art in the Age of Mechanical Reproduction," argued that making an exact copy of an object of art changed the entire meaning of the original work by placing it into a new context, we might now share the same concerns not just about art. but about all of experience.[16] In part, Benjamin's concern was that when it became possible to produce exact copies of artifacts, those copies, removed from their original contexts, not only devalued the originals from which the copies derived, but changed their meaning as well. In a similar sense, being able to place myself into an immersive simulation of a real experience, though it will allow the advantages of mass distribution and sharing of that experience, will inevitably dilute and devalue it. When my children shrug their shoulders at the appearance of an authentic dinosaur bone and turn instead to an augmented reality screen that shows an artist's conception of how the dinosaur that owned the bone might have looked while running, I believe that such devaluation of the real is exactly what has taken place. There's nothing very special about the thing itself because at a time of one's choosing, one can dial up an authentic 3D immersive experience that will have exactly the same effects on the senses. The context of the experience, in a world that is stitched together as a patchwork quilt of temporal and spatial moments that have been immortalized in silicon, slides away into insignificance. If Benjamin struggled with the implications of issues such as these for artistic reproduction, then the implications of reality reproduction are both mind-boggling and a part of a future for which we have not begun to prepare.

CHAPTER 8

SPACE AND TECHNOLOGY: THE MACHINE IN THE WORLD

W HEN WE ENTER AN IMMERSIVE virtual environment, we are essentially putting our minds inside a computer. For the most part, we are freeing ourselves from the physical world and putting ourselves under the control of a program that organizes what can be seen and heard, and monitors our movement responses with great accuracy to generate the grand illusion that we are somewhere other than where our body resides.

There's an entirely different way in which technology can permeate space, though, and the best way to characterize it is as a kind of inverse of the virtual "head-in-the-machine" metaphor. In the ubiquitous computing (Ubicomp) approach, the machine *becomes* the world. The late Marc Weiser, a research scientist at Xerox's Palo Alto Research Center and the first exponent of Ubicomp approaches to the relationship between environment and technology, saw this new approach as a natural evolution of the role of technology in our daily lives, and one that was much more in tune with human behavior than previous iterations of the human–machine relationship that began with the installation of gigantic room-sized mainframe computers and then slowly evolved to the more compact personal computers that sit on our desks.

In a heavily cited paper written with John Seely Brown in 1996, Weiser explains the transformational impact of Ubicomp using a simple and compelling metaphor: the inner office window. The window offers a subtle bidirectional view. The worker in the office can see some activity outside the door. A crowd of passing heads might signal an event such as the beginning of a meeting or lunch. A head bobbing repeatedly into view might let the occupant know that someone is outside waiting for an opportunity to visit. From the outside, the light shining from the office window signals the presence of an occupant. A brief glimpse could also inform the outsider of the nature of activity within the office. Is the occupant alone? Is she speaking on the telephone? Weiser and Seely-Brown described the function of the office window as "calm technology." It was calm because it did not force itself into the central view of parties on either side of the door, but it informed quietly, from the periphery, and helped to organize the behavior of both parties. In a paper that is both terse and profound, Weiser and Seely-Brown describe the main advantage of such technologies as being able to "make us feel at home," or in their terminology, they produce a feeling of "connectedness" by keeping us abreast of events that we need to know about, but without demanding taxing focal attention. Indeed, it may only be a small stretch to draw a parallel between the processes thought to be at play in ubiquitous computing approaches and those I mentioned earlier when describing the beneficial employment of casual, unfocused attention or "fascination," which seems to be a natural default means of paying attention to our environment in natural settings.[1]

When I was growing up in Toronto, I was often fascinated, like so many others, by a weather beacon that sat atop the Canada Life Building, a pretty Beaux Arts style structure that served as the headquarters of Canada's oldest insurance company. The beacon, which still operates today, juts from the top of the building and displays a continuous light display that codes a simple weather forecast. The beacon simply indicates the state of the skies (cloudy or clear), precipitation (blue for rain, white for snow), and whether conditions are changing or steady, whether temperatures are rising or falling. The weather beacon, though it only imparts a minimal amount of weather information, much of it also obvious from a quick glimpse at the sky, is a superb example of Ubicomp in action in the urban scene. Passersby who are familiar with the code can choose to glance upward or not; the beacon conveys a morsel of useful information from the edge of awareness.

When Weiser was formulating the central ideas of Ubicomp, it's difficult to imagine that he could have envisioned the explosion of embedded computing that now invades every corner of our environment. Writing in his popular blog *City of Sound,* urbanist Dan Hill's essay "Street as Platform" begins with a long description of all of the forms of information contained in the data cloud that envelops the average city street, much of it emerging from the quiet operation of networks of simple, decentralized sensors and processors. Such a cloud would include not only a mass of personalized data emanating from the phones of anyone on the street, but could also come from devices installed on the street to monitor traffic, business transactions, and temperatures (both outdoor and indoor), even including the ambient conditions inside any refrigerators located near the street, sidewalk load, the performance parameters of any cars located on the street, the state of the street's parking meters, or any one of a myriad other sources. The data cloud would include an amalgam of both public and private data, much of it networked and available to observers from remote locations. The data that pass from place to place on an average city street have become just as important as the steel and concrete that form the physical structures.[2]

In 1996, when Weiser and Seely Brown published their brief article outlining the philosophy of ubiquitous computing, it would have been difficult to have anticipated the incredible growth of one particular device—the one that has had the biggest impact of all on the way in which technology enters our lives: the smartphone.

When I arrived in Waterloo to take a professorship in 1991, I recall seeing some unassuming posters with strips of tear-off phone numbers plastered on the walls of my building. Now I wish I had taken a picture of them; at the time, it would not have been nearly as easy to do so as it is today. The posters were placed there by an upstart young company headed by Mike Lazaridis, a former student of the university, for a company that he was calling Research in Motion. The poster gave a short description of the basic idea of providing a computing platform housed in a tiny package that users could place in their pocket and carry with them. The company was looking for recruits. I remember standing and reading the advertisement and in one of my more stunning failures of imagination, I scratched my head and wondered who on Earth might find a use for such a weird, niche product. It sounded like one of the fancy scientific calculators used by engineering students, but with the capability to make phone calls. Lazaridis's company went on to produce the phenomenally successful BlackBerry smartphone, growing for a time to be one of Canada's largest high-tech companies, and when it went public in 1997, fueling the retirement savings plans of many lucky Canadians. At that time, the penetration of the smartphone in the global market was small. On a per capita basis, about 4 percent of the world's population owned smartphones. Contrast this with the 2014 figures that show that, globally, almost 70 percent of the population worldwide can access a smartphone and that the number of cellphones in use in North America almost equals its entire population.[3] In less than two decades, a device that was once the domain of wealthy early-adopters has become a basic accessory of life. Furthermore, the computing power of these devices exceeds by a wide margin the performance of computers that were available just a few decades ago. A single iPhone 5s, for example, which will soon be a model so antiquated that my teenaged children would be embarrassed to be seen with it in public, contains more memory and processing power than all of the computing equipment that powered the Apollo 11 mission to the moon. The same phone's graphic capabilities outstrip the performance of the famous Cray supercomputers of the 1970s by a factor of a thousand.[4]

Much has been made of the transformation of everyday life that has been fueled by the development of this remarkable technology; it is beyond the ambit of this book for me to recount very much of it. But there is one area, in particular, that is central to the role of the smartphone in the reorganization of everyday life with which everyone should be concerned. Buried within the guts of every smartphone is a tiny chip that by listening in on signals from a set of satellites originally launched into space by the United States Air Force lets the phone know where it is with an accuracy of, at worst, just a few meters. This kind of chip is available in the hobbyist market place for just a few dollars and for manufacturers of phones and similar technologies who buy in bulk, much less money than that.

It's hard to know where to even begin to describe the global repercussions of the advent of small, cheap, GPS sensors in mobile devices, but as our main focus must remain on the influence of technology on our relationship with built space, the obvious starting point seems to be to look at the pervasive influence of navigation technologies on wayfinding.

Young adult readers of this book will be hard-pressed to remember a time when it was not possible to solve the problem of being lost by pulling out their phone, finding their location on a small map by availing themselves of a signal sent from outer space to their phones, and then punching in their desired location. For confused travelers trying to make their way to an appointment in a strange city, this ability to shortcut all of the traditional methods of wayfinding used by human beings for thousands of years is a tremendous boon. Even for a casual tourist, the ability to call up a cellphone application and ask it for a list of nearby restaurants, perhaps complete with a compass readout that shows which way to go, the location-based technology that resides in our phones can be a time-saving marvel. Cellphone applications can also be used to annotate our surroundings for us. One application provides detailed historical and architectural information to the user that is cued by location. As we walk past an interesting building, our phone will pipe up with a brief story and some nuggets of trivia. Replacing the paper and pencil to-do list, our phones can also be programmed with geofences that remind us to do particular things when in particular places. For example, I've sometimes programmed my phone to remind me to pick up grocery items at the moment that it detects that I've left my office.

There is no doubt at all that our cellphones have both simplified certain problems that used to be hard to manage and made possible some new kinds of behavior that would otherwise be difficult. As with most of the technology that I've described in this book, I confess that I am an ardent fan and frequent user. But of course, there's a dark side. And I'm not talking here about the occasional lapses in GPS caused by signal or equipment failures or by out-of-date electronic maps; by now we've all had our fill (or so I hope) of stories of hapless travelers who find themselves driving into houses or lakes because they blindly followed the instructions from a GPS device while ignoring the evidence before their eyes. The more insidious risk of an overreliance on GPS is *related* to these kinds of mishaps, but it isn't quite the same thing.

First, imagine yourself in a world free of GPS applications in which you were required to manage your daily transactions with unfamiliar spaces old-school. If you are old enough to remember those days, then all the better. There are many different ways of navigating unfamiliar terrain, some that depend on our level of expertise and others that depend on the exigencies of the particular situation. If we stop to ask someone for directions, then the form they will take is likely to be a turn-by-turn description of a route, somewhat similar to the solution provided by a GPS device. Such a route-based description can work well for the immediate task at hand, but it's unlikely to tell us very much about the surroundings of our route or to convey any sense of the geometry of the places that we pass through on the way to our destination. A better way of comprehending our place—one that will not only get us safely to our destination, but also afford us with the possibility of further exploration and a fuller understanding of where we are—will be based on a map. Some of us (though a shrinking number, it seems) might resort to an actual paper map. The map, of course, is a scale reproduction of the patterns of streets surrounding us. One of the hard problems of using a map is being able to locate our current location; this often means that we must look closely at the landmarks that are presented on the map, look up at our surroundings, and accomplish a match between the two. Simply solving this problem has already given us much more information than the simple turn-by-turn description of a route. The sine qua non of superb navigation is the mental or "cognitive" map. This map is, in some ways, quite similar to the paper map that might still reside in the glove compartments of our cars (honestly, it's more likely that we'll find a map there than the gloves after which the compartment was originally named!). The most important difference between a cognitive map and a paper map is just that the former is composed of neurons, connections between neurons, and firing patterns of those neurons.

To be useful, it isn't critical that the geometric accuracy of our mental maps be perfect. In fact, they are likely to be distorted considerably, but as long as they preserve the relationships among the major landmarks in the map (that is, they have topological accuracy—think of something like a subway map), they will not only help us to get from point A to point B, but they will also give us a good idea of where to go if we also want to find points C, D, and E. Developing accurate cognitive maps of places requires some effort. We need to look around, pay attention to our surroundings, and engage in some effortful processing of the spatial relationships among the things that we see. In essence, we are trying to convert our views of things from ground level to a kind of omniscient bird's eye view of our surroundings. We're trying to imagine how things look from a perspective that we will never actually see: for most of us, this isn't easy.

What's interesting about the two different modes of wayfinding that I've described—one based on turn-by-turn descriptions of routes and the other based on a flexible cognitive map—is that each relies on very separate neural systems. Route-based learning resides mostly in an area of the brain called the striatum, which is a part of the basal ganglia. This collection of brain structures, buried deep beneath the surface of the cerebral cortex, was traditionally thought to be mostly involved in certain kinds of movement control. For example, some common movement disorders, such as Huntington's chorea and Parkinson's disease, result from damage involving structures in the basal ganglia. But we now know that parts of the basal ganglia, and especially the striatum, are involved in memory and learning processes, including the learning of routes.[5]

The hippocampus is a kind of ancient cortical tissue that is buried in the brain's temporal lobe. In groundbreaking work that won John O'Keefe a share of the Nobel Prize in Physiology or Medicine in 2014, he discovered that there were many cells in the hippocampus that were sensitive to location. These so-called place cells would increase in activity only when an experimental animal was in a particular location in the environment. O'Keefe's discovery of place cells set into motion a cottage industry in neuroscience in which investigators discovered ever more novel methods for recording the activity of hippocampal cells and for correlating their physiology with the behavior of animals, mostly laboratory rats. More recently, research using a variety of other types of neuroscience methods has shown that human beings have such cells along with probably all other mammals, and that these cells form the backbone of the human cognitive map.[6] The well-known studies of London taxi drivers by Eleanor Maguire, for example, have shown that the size of certain parts of the hippocampus is correlated with navigational expertise.[7] Still other experiments have shown that the hippocampus has a very desirable property for a structure charged with computing spatial information: it's beautifully malleable to experience. Not only do the physical properties of the hippocampus change radically in response to learning, but this is also one of the few areas of the brain where we are sure that new cells can actually be born during adulthood, overturning the centuries-old dogma that once we've reached early adulthood, all the brain cells we will ever have are with us and that, for the rest of our lives, brain changes are characterized mostly by a slow (hopefully) and inevitable process of cell death.

The relevance of these special properties of the hippocampus and their role in map learning comes from a consideration of the massive upsurge in our use of technology for wayfinding. By focusing on the blue dot of a phone map, rather than looking about at our surroundings and making the effort to form a genuine map, we are short-circuiting the processes that we've learned to use over previous millennia. As far as finding our way is concerned, we have become striatal stimulus–response machines, racing through time and space like feverish maze mice hunting for cheese.

Nobody has been more outspoken about the risks of this modern shift in behavior than McGill University neuroscientist, Veronique Bohbot. In a presentation at the 2010 meeting of the Society for Neuroscience, the largest annual congress of neuroscientists in the world, Bohbot made a splash by presenting data showing that elderly habitual GPS users not only had impoverished hippocampi compared to a control group who did not use technology to find their way, but that the tech-adopters also performed more poorly in a range of other cognitive tasks.[8] Bohbot warned that given what we know about the response of the hippocampus to experience and the general use-it-or-lose-it modus operandi of brain cells, habitual overuse of GPS to circumvent richer map-based interactions with place might actually trigger degenerative brain changes resembling those seen in dementing diseases like Alzheimer's disease. Indeed, cellular changes in the hippocampus are known to be an early harbinger of more serious degenerative pathologies.

As if the risk of brain damage isn't enough to make us worry about the longer-term effects of GPS, I have other kinds of concerns with an overreliance on location-finding technologies that are less tightly focused on wayfinding and have more to do with our general psychological relationship with place and how it is being affected by technology. One way to think about this is to consider the arguments of the philosopher of science, Albert Borgmann. In one of his early books, *Technology and the Character of Modern Life,* Borgmann describes what he calls the "device paradigm." The paradigm is a way of thinking about the relationship between "devices" and "focal things and practices."[9] A device, Borgmann says, is a piece of technology that presents us with a simple interface—an unassuming face—behind which hides the gadgetry that carries out a task that is beyond the reach of awareness. A simple example should make clear what Borgmann means by this and where he sees the risks. Many of us live and work in buildings that have heating and cooling systems of some kind. The unassuming interface of such a system is the thermostat, a very simple-looking device that ingeniously allows us to set the desired temperature of our surroundings. Using a simple mechanical switch and an element that measures temperature, the technology that hides behind the thermostat takes care of everything else for us. The wires that lead from the thermostat control the rest of the heating and cooling system. Variations in the operation of the systems required because of changes in the weather or our own habits (the forgotten open window, for example) are dealt with seamlessly and automatically. We need only touch the thermostat from time to time, remember to pay the utility bill, and occasionally have a technician service the system. To see how anyone could possibly imagine that the thermostat poses a societal risk, imagine an alternative. For me, it's as easy as remembering a delightful old relative of mine who lived alone in an isolated farmhouse in the middle of a windswept field in rural New Brunswick in Canada. Although she had a furnace, the main heat for her house came from an enormous wood-fed cast-iron stove that she had in her kitchen. To keep the house at the proper temperature, the dear lady had to periodically feed sticks of wood into the stove. She also had to remember to ask someone—her son, her grandchildren, or me when I visited—to be sure that she had a supply of wood nearby in the room that was adjacent to the kitchen. Not only this, but she had to have a good idea of the weather. When the forecast called for a cold snap, she knew that she would need more wood and that she would want to let the fire in the stove burn long into the night. Collectively, the activities and forethought required to heat her kitchen effectively—Borgmann would call these a focal practice—made a connection between my old friend, her community of family and acquaintances, and the greater environment. It should be clear that, although it might have made aspects of her life much easier to simply have a thermostat that she could nudge in the right direction from time to time, there would have been a loss, a disconnection, an isolating coldness to the substitution of a device for a focal practice.

Now, following the same kind of thinking for a cellphone that never lets you get lost, that provides you with an easily found blue dot to guide you like a perverse inversion of Diogenes' lamp leading us not to wisdom but slavishly to a preprogrammed destination, it shouldn't be hard to see that there's a similar kind of loss. In this case, the focal practice that has been cut off at the knees has to do not only with the social practices associated with wayfinding (cultural sharing of wayfinding knowledge, simple asking of directions), but also with our awareness and appreciation of the fabric, appearance, and meaning of place. In my book *You Are Here*, I described the strong coincidence that ancient wayfinding cultures like the Inuit in Canada, the traditional marine navigators of the South Pacific, and Australian aborigines not only relied on a panoply of ingenious skills to find their way, but they also had a feeling of intimate, reverent connection to their surroundings. The factor that underlies both navigational prowess and environmental stewardship is an attitude of careful, caring attention to everyday details, no matter how minute they might seem. It isn't impossible to imagine that wayfinding technology might somehow be co-opted to encourage that kind of caring attention, perhaps by means of the development of a playful relationship with technology that provides us with something more subtle, challenging, and provocative than a cartoon map of streets with a flashing arrow and a set of voice commands, but it's unlikely that a harried business traveler looking for the nearest Starbucks would want to have much to do with such a toy.

Can GPS-enabled phones really be so bad for us? Perhaps only in the same sense that television might be said to be bad for us. A device that permits us to sit on the couch for eight hours a day, watching mindless reality television shows or fatuous cartoons does just that: it *permits* maladaptive behavior, but it doesn't force it upon us. The same device that brings us mental pablum can also help us to be informed and educated. In the same manner, a smartphone can perhaps kill hippocampal cells and cut us off from attention to the details of our environment, but it can also get us to meetings on time, embolden us as travelers to go further without fear of the risks of disorientation, and in the hands of many types of emergency management professionals, probably even save lives.

In the case of mass media like television, the added value of the technology comes from those who work to provide valuable content. But what is the equivalent to high-quality content for a GPS-enabled cellphone? Is it possible to go beyond the blue dot and to use the capabilities of location-aware technology to enrich our relationship with place? The first place to look for applications that might encourage deeper, meditative, and playful encounters with place and space seems obvious: the world of gaming. One simple example of a geogame that has been around now since the inception of high-accuracy GPS signals in the year 2000 is geocaching. In this recreational pursuit, players look for secret treasures, or caches, that have been hidden by other players who have also entered the location of the cache along with a few clues in a public database available at a website. The caches may be solitary finds or they may be a part of a larger collection of caches that collectively consist of a puzzle or a narrative. It is difficult to estimate the number of geocachers worldwide, but this game, which is a little bit like tech-assisted orienteering, is one of the older and more robust forms of location-based gaming. The value of the game consists mostly of its encouragement of players to go out into the world, pay close attention to environmental details (often required to hunt down a carefully hidden cache), and to do some of the work that is often lacking in less mindful uses of GPS by expending effort to relate what is shown on the display of the technology that is being used to what is being seen in the environment.

There are many other types of location-based games that often contain puzzles, involve group activities, and sometimes culminate in exciting cat-and-mouse games requiring players to dash through interesting urban places to beat out competitors. But as compelling as these games might be, they very much represent a departure from real-life concerns—an attempt to turn the real world into a magic circle of game space. If imaginative content is to help to engage us with space rather than turn us away from it, then it must be somehow embedded in our everyday uses of space and technology.

At Yahoo Labs in Barcelona, Daniele Quercia's group has developed a GPS application that can find routes that are not based solely on a shortest-distance-to-destination algorithm, but instead on a series of aesthetic variables.[10] To build this application, Quercia's group first collected crowd-sourced data on the aesthetic values of urban viewpoints. To do this, they simply made available large numbers of images of urban locations and invited participants to rate the images for beauty, happiness, and quietness. Based on this feedback, the group went on to try to define the visual properties of urban scenes that were most likely to elicit strong impressions of those three aesthetic properties among viewers, and they applied their findings to an even larger set of images from the photo-sharing application Flickr. Having mapped some of the affective qualities of large areas of the cities of London and Boston, the final stage of development of the app was to build a wayfinding algorithm where users could not only specify their origin and their destination, but they could also request that the app find for them the happiest, quietest, or most beautiful route between the two points. The app is still under refinement, but early tests have shown that users responded positively to its use. Quercia's work, with its intention to undermine the simple shortest-route mindset of a user and encouraging them to spend a little bit of extra time for a more spatially enriching experience, holds some promise.

More on the bleeding edge of potential uses of GPS technology to subvert or extend both our ordinary use of wayfinding technology and our everyday understanding of space was the no longer available MATR app designed by the visionary interdisciplinary group Spurse, a creative design consultancy focused on transforming our understanding of systems ranging from communities and institutions to ecologies. MATR, which stands for Mobile Apparatus for Temporality Research, consisted of an app which tracked the location of the user on a specially constructed map containing much more than geographic features. The MATR map also included data relating to the historical climate, ancient geography, and human history of the locations for which it coded. Because these data included the history of a place over the long reach of time, MATR called them "deep time variables." This complicated amalgam of geographic and human data was mixed down to a sound that users of MATR could hear on headphones as they walked from place to place. The intention of the MATR app was much more ambitious than an application designed to help us treasure-seek for beauty and quietness. It was designed to provide an entirely new sensory channel grounded in both the technology of wayfinding and a deep set of spatial variables that reach into the vastness of ancient time.[11]

Although neither applications like MATR or Quercia's pleasure-seeking wayfinding app are likely to supersede our everyday use of our phones to avoid becoming lost, I like them both for the way that they provoke us to consider more carefully the impact of those everyday uses. Simply by virtue of providing an alternative to shortest-route algorithms and their mad dashes from point A to point B, both applications highlight some of what is being lost when we allow our native cognitive abilities to be consumed by a machine. They help raise us up to consider that the same device that might be used habitually to provide the spatial equivalent of a mindless celebrity game show is also capable of providing us with a riveting documentary that changes how we think about the world.

Anyone who habitually relies on a smartphone knows that the location-sensing capabilities of the device extend well beyond simple wayfinding applications. We also use our phones not just to tell us *where* we are, but also to provide information about *what* is around us. Almost every cellphone application will ask for access to your location information, whether the reason for wanting the information is obvious or not, and we are all too often willing to surrender this useful data point without more than a moment's reflection. There are at least three different ways that that information is used. First, and most obviously, knowing your location allows your phone to provide you with curated information about your surroundings. This is not simply a matter of telling you where you are on a map, but it could also involve letting you know what restaurants are nearby, where the closest cup of coffee is, and how highly rated by other users the shoestore in front of you might be. There's no question that having this kind of information at your fingertips can be useful, especially if you don't know the landscape very well. Second, many applications will allow you to contribute your own thoughts and feelings to the accumulated database of location-coded information. If you eat at a restaurant and enjoy yourself (or otherwise), you can rate the experience, write some words, and even contribute a photograph of your meal if you are so inclined. By donating information to a crowd-sourced repository of reviews, you are adding great value to the data that are generally available to users of these applications, the vast majority of which are provided free. Again, it seems as though one might have to have a severely dyspeptic disposition to find fault with a tool that allows the user access to such a treasure trove of valuable information at no cost other than the price of a phone and a data plan. I want to first allay any concern you might have that I'm going to tell you that using applications like *Yelp* or *Foodspotting* might damage your brain. As far as I know, it won't (though it might expand your waistline!). But I do have moderate concern with systems that are designed to provide instant online curation of the environment. I think that the real risk of these kinds of applications is that they can rob us of the raw experience of unmediated novelty. Along with most other animals, we have a kind of a sweet spot for novelty and complexity. What keeps us engaged with our environment is the promise that if we keep moving and looking, we will encounter things that will surprise and delight us. If we are led through the world by our phones, acting as a kind of information-rich range finder that advises us of what's ahead before we get there, then the devices are placing an extra layer between us and the pleasurable, unmediated, raw experience of the real. The only way that we can "stumble upon" something is while surfing the Internet. Real-world stumbling, with all of its serendipitous glory and energy, is becoming an anachronism.

But it's the third use of the mounds of location data that come from our phones that worries me the most. It only takes a second of reflection to see the too-good-to-be-trueness of a huge bounty of applications that provide useful information to us free. It's because what we are giving away must have value to someone else. Think of it this way. Imagine that we could go back in time to the grand days of the earliest department stores, places like London's Selfridge's or Le Bon Marché in Paris. The designers of those palaces of consumption were struggling to figure out how to get into the minds of shoppers—to make them happy and comfortable and excited about spending their money. Imagine offering those designers a tool that would allow them to peer inside the heads of the shoppers, to see where they went, what they looked at, how they described what they saw. Of course, they would leap at such an opportunity. Think of how, just a few years ago, it was common enough to see marketing surveyors standing in shopping malls with clipboards trying to stop shoppers to ask them a few questions. We don't see much of these people anymore because they're not needed; we are voluntarily pouring all of the data that marketers need into our phones and surrendering the right of ownership of our habits, movements, and thoughts to whomever will provide a simple phone app to us for free. In his pamphlet *The Epic Struggle of the Internet of Things,* science-fiction author and futurist Bruce Sterling makes a similar point, but in more flamboyant language.[12] Though he is talking about much more than our voluntary surrender of the details of our lives, he describes the major players—the Amazons, Googles, and Apples of the world—as being involved in a swashbuckling battle for world supremacy. As the entire physical workings of the world become mere proxies for their silicon incarnations in bits, the information of the world is the supreme asset and the last thing worth fighting for.

Once More, with Feeling

The new generation of mobile technology is capable of much more than simply tracking our movements and recording our purchasing habits. It can measure, in some cases quite directly, how we feel. When I conducted my world psychogeography experiments, exploring the relationships between different kinds of urban settings and human behavior, I used a simple wristwatch-style device to measure the state of my participants' autonomic nervous system using sensitive electrodes that unobtrusively recorded sweat gland responses. By correlating the ups and downs of their arousal levels with their location using a GPS-enabled smartphone, I was able to compile a stress map of an urban neighborhood based directly on fluctuations of the bodily state of my observers. It wasn't very surprising that visits to urban parks and gardens caused nice decreases in arousal, but the patterns of arousal also correlated with the types of street façades people encountered and the patterns of traffic and noise in the area. For a researcher like me, being able to visualize how the warp and woof of city streets influences one simple index of emotional arousal represents a tremendous advance. It isn't hard to imagine how such data might be used both to advance theory in urban psychology and also to address practical issues in city design. The sense of a planner, for example, that a street or a neighborhood might be producing unacceptable levels of stress, can now be supplemented with hard data that can quantify the size of the effect. I admit I feel nothing but excitement for developments in sensor technology that will make it easier and easier to directly sense people's emotional states while they are in naturalistic settings (the latest that I've seen is a method for manufacturing removable tattoos that contain embedded electronics for measuring sweat gland responses, heart rate, and even goosebumps.). But as the current rage in wearable technology continues, we will likely see many of the same kinds of tools in the consumer marketplace. Fitness accessories like the Fitbit or the Nike Fuelband allow users to measure their activity level, and in some cases, cardiac variables as well. A local start-up company in my hometown of Kitchener is designing a bracelet that will actually be able to peer into the blood of a user and analyze its content to extract metabolic indicators. The software licenses for such devices generally ask users to consent to the surrender of their data to the purveyors of such equipment. Just as with cellphone applications that monitor our movements and interests, we are just now on the thin edge of the wedge with regard to how such data might be used. For example, recently a woman in Canada has made headlines by submitting activity data from her Fitbit as evidence of loss of mobility in a personal injury suit.[13] Although in this case, the data were provided voluntarily, it seems inevitable that the day will come soon when such data will be subject to subpoena in law cases.

Apart from direct physiological measures, there are other ways of extracting geocoded indicators of affect from data that stream from our mobile devices. For example, the social media application Twitter, which has taken the world by storm by providing a free "microblogging" service by which users can share news, discoveries, insights, or photos of their lunches, can also be used to map the world's feelings.

Twitter feeds can be mined for emotional content using computational linguistics. So-called "sentiment analyses" have become big business as corporations can use them to measure attitudes of users about a particular product (by, say, analyzing the emotional content of all the tweets that contain the word *Starbucks*). But tweets can also be geo-coded, which means that they could be used to map the frequency of use of emotion words in different locations. Theoretically, it's possible to tag the location at which a tweet occurred with city block precision, but this depends on the privacy settings set by the user of the application. It's more common for tweets to be coded only to the home city of the tweeter. Nevertheless, the possibilities for using sentiment analysis, or even *intention* analysis, where text information is mined for clues as to what you plan to do *next*, will probably play an increasing role in the uses of social media to probe our inner states. With geographical variables added into the mix, this will make available to a wide range of commercial and institutional interests access to the emotional fabric of places.

Ground Control to Ground Control

Mobile phones have transformed our relationships with places. Using them, we carry in our hands a synthetic microcosm of our worlds that represents our location as a dot in a flat, stylized landscape and provides richly detailed curation of our surroundings. Many of us have become so enamored of this tiny recreation of the world that we sometimes pay far more attention to it than we do to the things that it represents. Our phones have opened up untold new possibilities for understanding the world, some of them good and others more worrisome, but because the phones are personal, the nodes of this great, connected network of devices still represent individual human beings.

The next frontier in the cybernetic transformation of space and place is not just focused on relationships between people, or even between people and the landscapes they inhabit. In the much-vaunted Internet of Things, places themselves are entirely penetrated by devices and sensors, still ostensibly in the service of human beings, but now with the central focus on the things themselves and their connections, rather than the flesh-and-blood actors who animate the scene. Many news media accounts might lead us to believe that what is new about the Internet of Things is that the appliances and gadgets of our lives will begin to talk to one another. Our carbon monoxide detectors will commune with our furnaces, knowing enough to shut things down when a lethal gas is detected in the air in our houses. Our wearable computers may talk to our appliances as well, so that when the fitness band we wear on our wrist detects that we have woken up, the coffee maker can be alerted to begin to put together the morning brew. In some ways, this might sound just like the responsive home, in which networks of sensors can come to learn our habits, adjust to them, and optimize their behavior so as to envelop us in a cocoon of knowing, feeling, and caring. What is different about the Internet of Things is a matter of scale. Rather than concerning themselves with the wiring together of a few household appliances to make our morning routines easier, massive corporations like Siemens and Microsoft are pushing hard to develop comprehensive systems that can do for an entire city what a Nest thermostat, which learns to turn down the heat when you leave your house and can talk to you via your phone, does for an individual abode. Indeed, entire cities such as Songdo City in Korea or Masdar in the United Arab Emirates, are beginning to spring from the ground complete with so-called smart city infrastructure. The utopian vision of the smart city is one in which the entire place is networked together to realize every possible efficiency. There is no traffic congestion, there are instant automated responses to emergencies, adaptive HVAC systems manage energy balances in the most efficient way possible, and other systems designed all the way down to the minute details of the lives of individual residents are there to take care of us. No detail is left unnoticed, no data point is left unarchived, and all of the complex machinery of the city from top to bottom is regulated using a set of complex algorithms designed to bring city function to its optimal state. It isn't hard to see the attraction of such a proposition. In a world groaning with overpopulation and a shrinking resource base, new problems in cities related to crowding alone can make us feel as though their solutions are so complicated as to be intractable to the human mind. What could be more appealing than the idea that we could wire everything together, attach it to a gleaming mission control headquarters filled with gargantuan computing devices that can comprehend and orchestrate all of the city's business, leaving us free to bask in the care of a city that works automatically and in the best possible way?

Nobody has done a better job of outlining the risks of smart city developments as they are currently envisioned than director of New York's design practice Urbanscale, Adam Greenfield. In his pamphlet *Against the Smart City*, Greenfield pores through the public relations material offered by the big players in the smart city market—Siemens, Microsoft, and Cisco—to try to deconstruct what these techno-giants might envision for our near future.[14] In his incisive analysis, he points out first that the schemes offered by these companies, at least according to the vision that they are trying to sell to the public and to the administrators of cities, appear to consist of one-size-fits-all system software designed to harness the collective power of the Internet of Things to optimize the function of a city using a set of all-encompassing algorithms. One problem with such an approach is that it assumes that the "problems" of the city are the same everywhere, and that what works in Songdo should also work in Paris, Berlin, or São Paulo. As Greenfield points out, offering a generic system to a collection of infinitely complicated and interesting cities ignores what matters most to the residents of a city: its culture, history, and personality. In this respect, he argues that it is no accident that the cities that have adopted smart city systems of the big players are entirely planned cities that have sprung from barren ground out of nothingness. They are really the epitome of Rem Koolhas' generic city concept, from which presumably they have drawn some inspiration. Just as worrisome is that, at least according to the glossy brochures and websites, companies like Siemens and Microsoft intend to monetize smart city systems by retaining complete control of their system software, keeping the software closed, and charging for its use. This kind of autocratic control of a city's systems from the top of the mountain seems incredibly dangerous. Just as juggernauts like Amazon and Apple have managed to exert massive control of the entertainment and book industries using powerful technology, companies that plan to dominate the market in the Internet of Everything stand to control, well, *Everything*.

From the psychological perspective, and thinking back to Borgmann, the most powerful question about smart city designs to me has to do with the impact that these designs will have on the behavior, feelings, and perhaps even the personalities of the residents of such designs. When we are coddled in a cozy envelope of caring that contains cars that drive themselves, baggage that knows when it has been stolen, pills that know when they have been swallowed, forks that know how fast we are eating, and diapers that know when they have been soiled, we may well feel cared for and safer than we've ever been, but the focal practices of our lives will be vastly diminished. When will we come face to face with the raw contingencies of life where we have to *decide* what to do based on incomplete evidence and judgment, like adults, rather than being passed from one automated system to another like swaddled infants? If it's really true that the use of a simple GPS system for wayfinding might cause changes in the organization of our brains, impoverishing neural systems through lack of use, then what effects would more comprehensive systems envisioned by smart city planners bring about? It's hard not to imagine that a feeling of helpless dependence and infantilization might be one of them.

Just as worrisome to me is the possibility that one-size-fits-all formulas imposed on cities from on high and designed for the generic human being might actually become self-fulfilling. Cushioned by a system of feedback loops that "protects" us from ourselves, it seems entirely possible—likely, in fact—that residents of a smart city, as envisioned by the current generation of smart city technology providers, might regress toward the mean. Though there will always be mavericks and dissidents who will struggle against the status quo and hack their environments to assert their own individuality, the corporate smart city will throw some prodigious new barriers in their way. In such a city, the trains might run on time, but there may be nowhere worth visiting.

If Greenfield is right in the optimistic conclusion to his short book, we may not have to worry about such a dystopian future. Though the terminology is different now and the technology has changed considerably, he points out that the bounty that is being offered by the big purveyors of the smart city is something that we've seen before, tried out, and dismissed as a failure. The modernist movement in city design, led most prominently by the Swiss architect Le Corbusier, proposed a similar kind of technocratic central control of the city, regulating urban activity in accord with scientific principles. Le Corbusier's Radiant City designs had some of the same hallmarks that are seen in the current smart city proposals—tight authoritarian control, scientifically designed affordances meant to optimize life for the fictional average person, and a completely mechanistic understanding of what cities are for and how they can thrive. Such designs, both at the level of individual buildings (Pruitt-Igoe, for instance) and at the scale of cities (Brasilia and Chandigarh) have been found wanting, to put things mildly, and perhaps most of all for the same misconstrual of human nature that Greenfield identifies in the smart city concept: we humans are not generic widgets who need only be inserted into the proper places in a larger machine. We are all different, messy, unpredictable, and disinclined to be tightly regimented in our daily lives.

It would be easy enough to completely dismiss the smart city concept on the grounds that it won't work well because it makes some of the wrong assumptions about both human nature and the nature of cities. This doesn't necessarily mean that major corporations like Siemens and Microsoft won't try quite hard to bring about their vision, if for no other reason than that a great deal of power and money is at stake. But it's also important to recognize that there may be other models for the use of smart, connected design using embedded environmental sensors, wearable technology, and mobile phones that will do much less harm and may bring great benefit. Just as I'm able to harvest data from small samples of participants in my experiments to learn about the psychological principles at play in urban and architectural design, some of the data that are generated by the technology we carry in our pockets, wear on our wrists, or install in our homes and cars could, if there was open, democratic access, provide citizens themselves with the tools that they need to quantify, understand, and act upon urban issues. Indeed, one of the objectives of my own work, especially when I conduct exhibitions or experiments designed to increase public engagement and understanding, is to make clear what is to be gained for all of us from the great boon in technology that is now available to measure our own behavior in the field. The keys, as with so many aspects of life, are to remain aware of the value of the information that we provide to the Internet when we use our devices, to ask the right questions about how those data are being used, and to refuse to meekly surrender control of our data to corporations, which, even considered in the most benign light, are not well situated to act as paternalistic overseers of the systems of a city, a building, or even the inside of our homes.

COMING HOME AGAIN

ABOUT TEN YEARS AFTER MY FIRST VISIT to Stonehenge, I went back with my father for a second time. Much had changed. For one thing, I was a brooding teenager. It was more difficult for me to stand beneath the stones and to imagine myself cowering under the massive legs of a giant. I still felt something—the presence of ancient powers, mystery, and wonder—but now I felt like pushing back, hiding my feelings by horsing around a little, skulking around as though this was not a big deal. But being there again in the company of my father meant as much to me as simply being there again. Mentally comparing the two experiences, separated by a time during which my age had more than doubled, gave me a way of calibrating the changes that had taken place in me. My inner life was more complicated. The stones still spoke to me, but now in a more analytical language. Like my father, my mental spaces were now as much filled with questions about the meaning of the stones and their methods of construction as they were with head-raising awe and shivering fear. I was starting to occupy the middle-ground of an adult responding to built designs: on the one hand still being subject to the subtle, unconscious influences of the geometry and surfaces of a building on my movements and feelings, but on the other hand adopting a more formal, distant, analytical stance—standing outside my own experience in a way, using our miraculous ability to be aware of ourselves. I peered at my father from the corner of my eye and I tried to imitate his movements and postures, narrowing my eyes to calculate sizes and shapes, trying to conjure images of the legions of struggling workers lifting and pushing these stones into place.

The place I occupied that day, that bicameral middle ground between the raw experience of a place and the outsider's analytic assessment of it, is one that I've never since left. It points directly to the most important theme of this book. Whether we like it or even realize it, places envelop us in feelings, direct our movements, change our opinions and our decisions, and maybe even sometimes lead us to sublime, religious experiences The story of how the forms of the built environment exert these effects is older than civilization, but still far from complete. The science behind these relationships is beginning to emerge. Major stakeholders are paying close attention to this science; this, combined with a remarkable revolution in the technologies that we can use to monitor behavior and build environments that are responsive to us, poise us to enter a new age of person–environment relationships.

There seems no limit to the variety of different approaches to understanding how these relationships between buildings and people work. We can understand buildings themselves as pieces of art, political statements, cultural artifacts, or simply as machines—generic containers for the conduct of human lives—and it is easy to find good examples of all such approaches in the voluminous history of architectural ideas. In my own work, I've tried to adopt a scientific approach grounded in the basic facts of psychology and neuroscience. But even as I try to apply sweeping reductive principles to the manner in which buildings and cities affect us, I can still sometimes feel the trembling legs of the six-year-old version of me, reminding me that reducing the meaning of a piece of architecture to a set of equations will always cause us to miss out on answers to some of the most important questions. I feel as though I'm standing astride a great divide: with one foot on the side where I'm captivated by the new generation of tools at my disposal that I can use to help to bring psychological science to design and the other foot on the side that is suspicious not only of the misapplication of such tools, but also of the kind of world that might result if such methods are applied rigorously and without limit. Whether you live and breathe the dilemma as I do, or whether for you it's more like a faint rumbling of a distant thunderstorm—both exhilarating and a little bit frightening—you stand astride the same divide that I do. We all know that something big is coming, in some respects it's already here, and we aren't sure how to respond to it.

Some time ago, I was asked to give a public presentation to an audience composed mostly of designers and architects. The presentation took the form of a discussion between me and a prominent Vancouver architect, moderated by another architect and described in the promotional literature as a "collision between neuroscience and architecture," a characterization that left me feeling slightly nervous. I had never felt that my own small contributions to understanding how we might design better buildings could collide with anything at all, least of all architecture. Perhaps naïvely, I had always felt that my role in the real world of design was to help to point the way toward evidence-based principles that could be extracted from the data from my experiments. I was only here to help make better buildings. The discussion went as such discussions often do: the architect described his creations and I described my science. But during the freewheeling discussion following our formal talks, some of the impact points of the collision began to emerge, helping me to understand the anxieties of an architect being confronted with scientific arguments about the connections between design and human psychology. It boiled down to questions of freedom. If I do an experiment that shows conclusively that round doors cause people to become less stressed, secrete less cortisol, and therefore be at lower risk for the development of cardiovascular disease, so the argument goes, then a zealous administrative body—say a municipal government charged with the maintenance of up-to-date building codes—might then decree that forever more, all buildings in that city shall have round doors. Although this particular example might seem slightly frivolous (though I wouldn't be that surprised if round hobbit-style doors actually *did* make people feel good), it highlights the dilemma.

It seems a risky course to so scientize design that the creative vision of architects is force-fed into a reductive sausage grinder that can only produce quasi-Corbusian designs of the kind that we've already tried and found wanting. Nevertheless, allowing an architect to have unfettered access to a fecund imagination untroubled by the psychological realities of what seems to work in a building also seems unwise. But to somehow suggest that we need to occupy a middle ground, where sensible design research is acknowledged, but the creativity of great design, just as impossible to capture in a laboratory flask as it would be in a genie's magic lamp, is given full play seems both a wishy-washy conclusion and an impossible one to achieve. Is there a third way?

Any way out of the collision between science and architecture would need to find a way to skirt around authoritarian and paternalistic solutions in which the principles of science alone would dictate inflexible solutions to city building. A fully contextualized application of such principles should give us the tools to understand how the design of built environments influences human behavior, and how these influences have been shaped both by the ancient wiring of our nervous systems and by the events of history—the redefinition of the role of focal attention on productivity as a virtue or even a prerequisite for modern life would be one good example of such a historical influence. The framework is not meant to be prescriptive, as I don't believe that there is any one-size-fits-all solution to architectural or urban design, any more than I believe that a top-down monolithic application of controlling technologies in settings of any scale—from the inside of a home to an entire city—can tell us how we should build. I *do* believe, however, that the application of scientific principles derived from careful cognitive science and neuroscience can both help us to see how systems of buildings are working now, and to make some intelligent predictions about how changes in the systems themselves might influence our behavior. But there is, and always should be, a separation between such predictions and any city-building prescriptions that might arise from a source of authority.

At the same time, there is little merit in allowing architects unfettered freedom to design any kind of structure that appeals to their imagination (nor do I really think this ever happens in real life). Others have pointed out the disconnect that can take place between the vision of an architect and the opinions of the users of the finished product. To put it bluntly, the preferences of ordinary people are often considerably at odds with the aesthetic judgments of designers. One can certainly make the argument that this gap is in part a matter of education; this is a point of view that I've had expressed to me with considerable fervor by quite a few architects, and it must be taken into account. A building may be a legitimate artistic creation for an architect, but unlike a painting, a movie, or a sculpture, the finished building must be capable of playing a useful and positive role in the lives of its users on a daily basis and over the lifespan of the construction. The architect has a public responsibility to care whether a building works well for its purpose and whether it makes a net positive contribution to the built landscape; psychological analysis and experimentation can help them to fulfill this responsibility.

But the third part of the system must involve the people—the users themselves. If we acquiesce to bad design, wallow in a shoulder-shrugging apathy, and suppose that the forces at play in the construction of our environments are so powerful, so authoritarian, and so beyond our simple understanding, then we will inherit the places that we deserve. Armed with understanding—and I hope that my writing here has made some small contribution to this—any intelligent, well-informed citizen should stand ready to enter the fray, offer an opinion, and contribute his or her own vision to the debate about how our built environment should unfold. And this is one area where our new hyperconnectedness, brought about by the Internet and mobile technology, can help tremendously.

The widespread availability of technology that makes it possible for any of us to collect location-based data on our own responses to our environment, up to and including the responses of our own bodily physiology, for all of its risks, offers great hope for citizen-based contributions to the effort to build better places. More than at any time in our past, there exist mountains of data describing where we go and how we feel while we are there. It's a rare cellphone application that does not offer us the opportunity to add a geotag to our data—reviews, photographs, walking or driving patterns, heart rates, accelerometry, and even sometimes body temperature and arousal levels. Although many applications send individual data to a central database owned by the company that provides the software, and those aggregated data are not available to the public, some apps at least allow users to look at their own data, and a few are set up so that we can compare our own stats with those of other people. In addition to this, a widespread movement for "open data" is gaining steam, and encouraging municipalities, states, and countries to make available to the public much useful data related to patterns of activity, traffic and economics. Theoretically, such data could constitute an extremely useful tool for the democratization of city design. Access to this new form of information, critical as it is to understanding how places work, should not only be easily available to everyone, but the basic tools for understanding how it can be used and what it can tell us should be available for all. Data science should be taught in our schools. Discourse in how cities work couched in visualizations built from big data is becoming so important that the basics should be included in the public educational curriculum, just as civics has been now for generations. And, as architectural theorist and historian Sarah Goldhagen has argued, so should architectural history and design. A curriculum that has a place for such things as fine arts and literature, important as they are for understanding the human condition, has no business ignoring a practice that has been with us for as long as we have been human, that surrounds most people on the planet for most of every single day of their lives, and that has such clear and profound effects on everything that we do, feel, and think. But knowledge alone is not enough. We must also act.

For reasons that are amply clear, there are a lot of data that will be difficult if not impossible to wrest from the hands of those who collect and store it—there's a great deal of money and power at stake. But set against this the fact that many grassroots organizations, particularly those associated with "Maker Movements"—groups intent on constructing devices themselves rather than relying on commercially available wares—are working hard to help people build their own tools for both the collection and the analysis of information that could help to flesh out our psychogeographic understanding of built space. Though such data may be messy and lack the clinical archival quality that can be acquired using formal experiments of the kind that I've described in this book, they will constitute useful starting points for the analysis of existing patterns, debate, and dialog with the governing bodies that dictate what can be built and where it can go. Such crowd-sourced, grassroots efforts have the great advantage that they help to make citizens themselves active participants in the processes that eventually lead to built designs.

None of this is to suggest that making ordinary citizens active participants in the processes that lead to new buildings will be an easy thing—anyone who has attended a city council meeting to voice an opinion about a proposed development will know that to suggest that stakeholders are always poised to listen to the public is the attitude of a naïve Pollyanna. When money is at stake, the game always gets rough. Nor am I suggesting that with crowd-sourced data gathering, the people can take matters of design entirely into their own hands. Though there is much we can do to contribute to the public dialog about the design of the built world, very few of us will be able to acquire the expertise that is needed to design a building ourselves. We must work in partnership with those who do have such expertise, and the best way to achieve such a partnership is to find a common language. The best ally for policymakers, planners, designers, and architects will be a well-informed public that not only understands how to listen to their senses and interpret what they hear in light of what is known about how buildings affect us, but also stands ready to contribute to that knowledge by carefully monitoring their own feelings as they explore the built world. In the psychogeographic workshops that I've run throughout the world, one of the most heartening things that I've learned is that, with the right tools, this is a skill that develops quickly and rewards participants with a sharper and more nuanced understanding of how the settings of the world affect them.

Head in the Clouds, Feet on the Ground

Apart from describing how the urban environment influences our psychological state—information that you can use to organize your own psychogeographic life and to perhaps contribute to the ongoing debate about how to build the best cities, neighborhoods, and places—the other main message of this book has to do with the implications of emerging technology for our understanding of space, place and ourselves. The repercussions of the coming changes will be both wide and deep; they are already well under way. Many words have been spilled about the influence of the connected world on our ability to pay attention and to remember, and the manner in which instant connectedness with others changes the nature of social relationships and politics, but here I am talking about something different.

In some ways, the story is as old as the earliest attempts to forge human nature with the arrangement of massive stones at Stonehenge or Göbekli Tepe. The methods we use now, still in the service of the same ends, are new and startling. What once might have taken years of effort to move earth and boulders can now be effected with the flipping of electronic switches and the bending of light rays. But at both ends of a continuum of time that stretches from antiquity until the present day, there is one common element. When we gained self-awareness, thereby fracturing the world forever more into an introspective parade of sensations, thoughts, and feelings, and an external world of physics and matter, we gained immense mental riches for ourselves but we lost the mute, mindless happiness of an animal that doesn't know that its life will end. It is that painful awareness of our own finitude, not always foremost in our thoughts but never very far below the surface of our daily activities, that lends architecture its potency. We use it to convince ourselves that though we may shrug off our bodies, some part of us will live on in the things that we build. We use it as a means of huddling for protection from the power of greater forces, just as we hid beneath the legs of our parents for protection from the things we feared. We use it to help us to look upward to the heavens, trying to break through the bounds of our bodies to achieve a mystical union with the universe and experience feelings of awe in the dissolution of the ordinary trappings of space and time.

Now we stand on the threshold of a time when we can enmesh ourselves in protective networks of sensors and actuators that watch over us and keep us safe. Part of the appeal of such smart designs might be that they help us to confront the enormous complexities of modern life in crowded cities, but another part must certainly be our desire to bury the bothersome details of keeping our lives running smoothly beneath a shiny technological interface that learns our behavior, gets to know what we need, and takes care of us like a meticulously attentive parent.

Now we stand on the threshold of a time when any of us will be able to enshroud our bodies in technology—head-mounted displays, cyber-gloves, augmented reality goggles—that will allow us to throw our precious self-awareness anywhere, to embody new shapes and sizes, and to experience any setting with high-resolution, surroundsound authenticity so convincing that we will easily achieve the holy grail of a simulated experience—true presence. It would take only the most miserable pessimist to see no good in this. We can enjoy enriched experiences for education, entertainment, and self-understanding. But at the same time, we run the risk of cheapening the real by blurring the distinctions between the precious, unique, fleeting, authentic experiences of our lives with convincing, easily duplicated facsimiles. It's hard not to think that this will take a metaphysical toll on us. It's hard not to believe that the easy availability of such technology will come to change completely our everyday understanding of what is real and what is not real. By throwing off the yoke of simple embodiment within the physical space of our bodies, as it has been for thousands of years of human evolution, we have begun to enter a new phase of civilization. Indeed, those of us who spend much of our lives suspended somewhere between our immediate physical surroundings and the play of messages, sounds, and images that arrive at the mobile phone in our hands, are already halfway there, perhaps without realizing it.

All of this began thousands of years ago with the construction of the first wall. That wall might have been there to keep some things outside and others in, but it also had the effect of changing the world's geometry in a deliberate effort to change what could be seen and experienced from either side of the divide. Though our methods might have advanced well beyond the imagination of the sweating laborers who lifted one stone upon another to make ancient temples, the underlying intentions are much the same.

For some time now, I've dreamed of taking my young son to visit Stonehenge. He's around the same age now as I was when I first visited the site with my father. I know that things are very different now— tourists are driven to the site in special buses, corralled by walkways and fences, and their visits are carefully timed. A motorway roars past the site at a distance that's a little too close for comfort. But even though he can never have what I had, standing alone with his father in the early morning chill breezes of desolate Salisbury Plain, I'd like to get him as close to those old stones as I can, to see if I can share with him any part of what I felt. There's also a beautifully crafted simulation of the Stonehenge site designed by the Institute for Digital Intermedia Arts at Ball State University in Muncie, Indiana.[1] I'm putting off sharing the idea of my dream adventure with him for now because I know that within minutes of my mentioning it to him he will find the website, download the simulation, and he'll be there, at virtual Stonehenge. It may be nothing more than a pipedream, but I still hope that I can help him feel the power of being there.

Notes

Introduction

1   It's pretty easy to find further information about Göbekli Tepe. A great starting point is the *New Yorker* article by Elif Batuman, "The Sanctuary," found in the December 19, 2011 issue of the *New Yorker*. Available at: http://www.newyorker.com/magazine/2011/12/19/the-sanctuary

2   I'll discuss these findings more fully in Chapter 6, Spaces of Awe, but if you're curious now, you can look up the paper by Stanford professor Melanie Rudd and colleagues, titled "Awe Expands People's Perception of Time, Alters Decision-Making, and Enhances Well-Being," in *Psychological Sciences*, Volume 23(10), pages 1130–1136.

3   In addition to his scientific publications, Antonio Damasio describes his theories and findings with admirable clarity in a series of books written for the general reader. The best exposition of the findings I have mentioned can be found in *Descartes' Error: Emotion, Reason, and the Human Brain*, (Putnam Publishing, New York, 1994).

4   Rizzolatti's original account of the discovery of mirror neurons, along with much of his subsequent work, is described in the technical review paper titled "The Mirror-Neuron System" in the journal *Annual Review of Neuroscience*, 2004, Volume 27, pages 169–192. With Corrado Sinigaglia, he has also written an account for the general reader in *Mirrors in the Brain: How Our Minds Share Actions and Emotions* (Oxford University Press, UK, 2008).

5   The rubber hand illusion was first reported by Botvininck and Cohen in a paper titled "Rubber Hands 'Feel' Touch that Eyes See," published in *Nature* (1998, Volume 391, page 756). It has been replicated dozens of times in experiments designed to explore issues of embodiment.

6   Henrik Ehrsson of the Karolinska Institutet in Stockholm first reported experimentally induced out-of-body experiences in a paper titled "The experimental induction of out-of-body experiences," in *Science* (2007, Volume 317, page 1048). The phenomenon has been repeated many times in many laboratories, including my own where we use the demonstration to interest students in issues related to embodiment.

7   A technical account of remapping of space using pointers is provided by Longo and Lourenco of the University of Chicago in a paper titled "On the nature of near space: Effects of tool use and the transition to far space," in *Neuropsychologia* (2006, Volume 44, pages 977–981).

8   Amy Cuddy's fascinating and popular TED talk can be found at http://www.ted.com/talks/amy_cuddy_your_body_language_shapes_who_you_are?language=en A technical paper describing some of the findings she discusses in the talk can be found in a paper titled "Power Posing: Brief Nonverbal Displays Affect Neuroendocrine

Levels and Risk Tolerance," in *Psychological Science* (2010, Volume 21, pages 1363–1368).

9 Maarten Bos and Amy Cuddy describe the effects of use of electronic devices of varying size on power postures and, through this our behavior in a paper titled "iPosture: The Size of Electronic Consumer Devices Affects Our Behavior," in *Harvard Business School Working Paper* (2013, No. 13-097). The paper is available at: http://www.hbs.edu/faculty/Pages/item.aspx?num=44857

10 Results showing the effects of holding a warm drink on social behavior are reported by University of Toronto researchers Chen-Bo Zhong and Geoffrey Leonardelli in a paper titled "Cold and Lonely: Does Social Exclusion Literally Feel Cold?" in *Psychological Science*, 2008, Volume 19, pages 838–842.

11 Joanne Wood of the University of Waterloo, along with her students, reported the effect of unstable seating on relationship preferences in an article with D.R. Kille and M. Forest titled "Tall, Dark and Stable: Embodiment Motivates Mate Selection," in *Psychological Science* (2013, Volume 24, pages 112–114).

12 John Locke's fascinating book *Eavesdropping: An Intimate History* (Oxford University Press, New York, 2010) describes the history of the wall and its psychological impact.

13 The "Internet of Things" is a term used to describe interconnected networks of electronic devices that optimize and simplify the human use of built environments.

14 Joseph Paradiso of MIT's Media Lab makes this provocative statement about the future of responsive environments in an interview with Sarah Wesseler published in the blog *ArchDaily*. Available at: http://www.archdaily.com/495549/when-buildings-react-an-interview-with-mit-media-lab-s-joseph-paradiso/.

Chapter 1

1 Roger Ulrich's groundbreaking study of the influence of views of nature on surgical recovery was first published in an article titled "View Through a Window May Influence Recovery from Surgery," *Science* (1984, Volume 224, pages 420–421).

2 You can find out more than you probably want to know about habitat selection in warblers in an article by Brown University biologist Jeffrey Parrish titled "Effects of Needle Architecture on Warbler Habitat Selection in a Coastal Spruce Forest," in the journal *Ecology* (1995, Volume 76, pages 1813–1820).

3 The Urban Dictionary is a whimsical crowd-sourced lexicon of current slang and euphemism. Available at: http://www.urbandictionary.com/

4 A description of habitat selection in laboratory environments for the mighty Manini may be found in Sydney, Australia biologist Peter Sale's account titled "Pertinent Stimuli for Habitat Selection by the Juvenile Manini, *Acanthurus Triostegus Sandvicensis*," in the journal *Ecology* (1969, Volume 50, pages 616–623).

5 The fascinating pole-dancing studies of Anolis lizards are described in an article by Kiester, Gorman and Arroyo in a paper titled "Habitat Selection of Three Species of Anolis Lizards," in the journal *Ecology* (1975, volume 56, pages 220–225).

6 Jay Appleton's book, *The Experience of Landscape,* has influenced a generation of landscape architects with its broad sweep and theoretical power. (Wiley, London, 1975).

7   Grant Hildebrand is the preeminent authority on the use of prospect and refuge in Frank Lloyd Wright's architecture. His book *The Wright Space: Pattern and Meaning in Frank Lloyd Wright's Houses* (University of Washington Press, Seattle, WA, 1991) is a fascinating and accessible read.

8   Jan Wiener and Gerald Franz describe experiments in generic art gallery spaces in virtual reality that suggest the power of prospect and refuge to predict preference in a chapter titled "Isovists As a Means to Predict Spatial Experience and Behavior," in the conference proceedings *Spatial Cognition IV*, C. Freksa, M. Knauff, B. Krieg-Brückner Bernhard Nebel and T. Barkowsky, eds. (Springer-Verlag, Berlin, pages 42–57, 2005).

9   The first complete description of the idea that people prefer savannahlike environments can be found in the interesting paper by Judith Heerwagen and Gordon Orians, in their chapter "Humans, Habitats, and Aesthetics," S. R. Kellert and E. O. Wilson, eds., *The Biophilia Hypothesis* (Island, Washington, D.C., 1993, pages 138–172).

10 Cross-cultural preferences for savannahlike environments were described by John Falk and John Balling in an article titled "Evolutionary Influence on Human Landscape Preference," in *Environment and Behavior* (2010, Volume 42, pages 479–493).

11 Studies of patterns of eye movements while participants looked at scenes of nature are described by Rita Berto and colleagues in a technical article titled "Do Eye Movements Measured Across High and Low Fascination Photographs Differ? Addressing Kaplan's Fascination Hypothesis," in the *Journal of Environmental Psychology* (2008, Volume 28, pages 185–191).

12 Rachel and Stephen Kaplan's magnificent book *The Experience of Nature: A Psychological Perspective* (Cambridge University Press, Cambridge, UK, 1989) is required reading for all environmental psychologists.

13 Frances Kuo and William Sullivan describe the fascinating relationship between urban crime and urban vegetation in an article titled "Environment and Crime in the Inner City: Does Vegetation Reduce Crime?" in *Environment and Behavior* (2001, Volume 33, pages 343–367).

14 For a fascinating but somewhat technical account of fractals in Pollock paintings see the article by Richard Taylor and his colleagues titled "Perceptual and Physiological Responses to Jackson Pollock's Fractals," in the journal *Frontiers in Human Neuroscience* (2011, Volume 5, article 60, pages 1–13).

15 Evidence for a fractal sweet spot for human preference in landscapes is described in the article by Caroline Hagerhall, Terry Purcell, and Richard Taylor in a paper titled "Fractal Dimension of Landscape Silhouette Outlines as a Predictor of Landscape Preference," in the *Journal of Environmental Psychology* (2004, Volume 24, pages 247–255).

16 Valtchanov's main findings are not yet published in the peer-reviewed literature but a compendious summary can be found in his thesis available at: https://uwspace. uwaterloo.ca/bitstream/handle/10012/7938/Valtchanov_Deltcho.pdf?sequence=1

17 Mary Potter of MIT has made a life work of understanding the psychological mechanisms of rapid scene perception. One of the first papers in this area was published by Potter and colleague Ellen Levy titled "Recognition Memory for a Rapid Sequence of Pictures," in the *Journal of Experimental Psychology* (1969, Volume 81, pages 10–15).

18   Irving Biederman and Edward Vessel have written a beautiful article for general
      audiences titled "Perceptual Pleasure and the Brain," describing their work on the PPA
      and aesthetic preference in *American Scientist*, (2006, Volume 94, pages 249–255).

19   Some of my work with student Deltcho Valtchanov on the restorative effect in virtual
      environments is described in an article titled "Restorative Effects of Virtual Nature
      Settings," in the journal *Cyberpsychology, Behavior, and Social Networking* (2010,
      Volume 13, pages 503–512).

20   Hunter Hoffman and colleagues describe the use of virtual reality for control of dental
      pain in an article titled "The Effectiveness of Virtual Reality for Dental Pain Control:
      A Case Study," in the journal *Cyberpsychology and Behavior* (2004, Volume 4, pages
      527–535).

21   Peter Kahn has written several good books about the relationship between the
      emergence of technology and our loss of contact with nature. One of the best of these
      is his book *Technological Nature: Adaptation and the Future of Human Life* (MIT Press,
      Cambridge, MA, 2011).

22   Elizabeth Thomas's account of life with the Bush People can be found in *The Harmless
      People* (Knopf, New York, 1959).

23   Lewis Mumford, *The City in History: Its Origins, Its Transformations and Its Prospects*
      (Harcourt, Brace and World, New York, 1961).

24   Jonathan Crary, *Suspensions of Perception: Attention, Spectacle, and Modern Culture*
      (MIT Press, Cambridge, MA, 2001).

25   The Canadian media theorist Marshall McLuhan revolutionized our understanding
      of the impact of media on communication. His best-known work is *Understanding
      Media: The extensions of man*, McGraw-Hill: Toronto, 1964.

Chapter 2

1   Beesley quote from an interview with Fran Schechter for *NOW* magazine (2010,
      Available at: https://nowtoronto.com/art-and-books/features/art-as-organism/)

2   Philip Beesley's curriculum vitae may be found at: http://philipbeesleyarchitect.com/
      about/14K24_PB_CV.pdf

3   One of the early and most influential accounts of rapid scene recognition may be
      found in Mary Potter's landmark 1969 paper titled "Recognition Memory for a Rapid
      Sequence of Pictures," published in the *Journal of Experimental Psychology*, (1969,
      Volume 81, pages 10–15).

4   Fritz Heider's classic study with Marianne Simmel was published in a paper titled
      "An Experimental Study of Apparent Behavior," in 1944 in the *American Journal of
      Psychology* (Volume 57, pages 243–259). The video I described is easy to find online,
      for example at: https://www.youtube.com/watch?v=76p64j3H1Ng

5   Albert Michotte's work on causality is found in his beautiful book *The Perception of
      Causality* (Methuen, Andover, MA, 1962).

6   I heard the story of the hoarder and the wet containers at a workshop held in Toronto
      in 2012 by the renowned psychologist David Tolin. His book, written with Randy
      Frost and Gail Steketee titled *Buried in Treasures* (Oxford University Press, London,
      2007), provides a fascinating overview of hoarding disorder.

7   Edgar Allan Poe's short story *The Fall of the House of Usher* is in the public domain and so is easy to find online. Many fine paperback reproductions also exist.

8   Many beautiful photographs of Malian mud architecture, including some traditional organic home spaces, can be found in the website Atlas Obscura. Available at: http://www.atlasobscura.com/articles/mud-masons-of-mali

9   Witold Rybczynski's fascinating book *Home: A Short History of an Idea* (Viking, New York, 1986) is filled with rewarding insights about the architecture of home spaces.

10 Peter Ward's book *A History of Domestic Spaces* (UBC Press, Vancouver, 1999) focuses on Canadian history, but many of his insights generalize to at least the North American context.

11 Hermann Muthesius's beautiful book *The English House* was published in German in 1902 and finally translated into English in a beautiful box set edition (Frances Lincoln, London, 2006).

12 Sarah Susanka's "Not So Big" series of books have influenced many to consider the functionality of a home rather than its brute size. The original book in the series, and the one most relevant to discussion in this book by Susanka and Kira Oblensky is *The Not So Big House: A Blueprint for the Way We Really Live* (Taunton Press, Newton CT, 2009).

13 The quote comes from page 15 of the English translation of Gaston Bachelard's influential book *The Poetics of Space* (Beacon Press, Boston, 1994).

14 Cicero described his so-called method of loci in *De Oratore,* which can be found in print in *Cicero Rhetorica.* Volume I (De Oratore), edited by A. S. Wilkins (Clarendon Press Oxford Classical Texts, Oxford, UK, 1964).

15 Gabriel Radvansky's experiments on working memory and doorways appears in an article by Radvansky and colleagues Krawietz and Tamplin titled "Walking Through Doorways Causes Forgetting: Further Explorations," in *The Quarterly Journal of Experimental Psychology* (2011, Volume 64, pages 1632–1645).

16 Minkowska's work with children's drawings is described in an article written by her husband, Eugène Minkowska, titled "Children's Drawings in the Work of F. Minkowska" in the *Annals of Medical Psychology* (1952, Volume 110, pages 711–714).

17 Carl Jung's own description of the tower can be found in his autobiographical book *Memories, Dreams, Reflections*, published in English by Vintage Press, London, 1989.

18 The quote from Jung comes from *Memories, Dreams, Reflections* (Vintage Press, London, 1989. page 212).

19 The complete report of the Pew Research Center on American Mobility can be found at: http://www.pewsocialtrends.org/files/2011/04/American-Mobility-Report-updated-12-29-08.pdf

20 Oscar Newman's influential book *Defensible Space* was published by Macmillan, London, 1972. Many of the important ideas from the book can be found in the monograph *Creating Defensible Space*, available at: http://www.huduser.org/publications/pdf/def.pdf

21 Aisha Dasgupta's work with the BMW-Guggenheim lab is described in an article by Neha Tirani titled "In Mumbai, Privacy Is Hard to Come By," in the *New York*

Times on January 2, 2013 Available at: http://india.blogs.nytimes.com/2013/01/02/
in-mumbai-privacy-is-hard-to-come-by/?_r=0

22 Nicholas Negroponte's vision for responsive architecture appears first in his book *The Architecture Machine: Toward a More Human Environment* (MIT Press, Cambridge, 1973).

23 A profile of Dr. Dan Vogel, including the quote used here, can be found at https://uwaterloo.ca/stories/bringing-science-fiction-home

24 The essay by Walter Benjamin was first published in 1936 in French in the journal *Zeitschrift für Sozialforschung*, Volume 5, pages 40–68. An English translation can be found at: http://www.marxists.org/reference/subject/philosophy/works/ge/benjamin.htm

25 Adam Scharr has written a fascinating account of the influence of home on Heidegger's work in *Heidegger's Hut* (MIT Press, Cambridge, 2006). The quote from his son, Hermann, comes from a BBC television documentary titled *Human, All Too Human*, which first appeared in 1999.

Chapter 3

1 Jack Katz, *Seductions of Crime: Moral and Sensual Attractions in Doing Evil* (Basic Book, New York, 1990).

2 The *Chromo11* website can be found at: http://www.chromo11.com/

3 Brendan Walker's *The Taxonomy of Thrill* was published by Aerial Publishing, London, 2005.

4 Rem Koolhaas's provocative book *Delirious New York: A Retroactive Manifesto for Manhattan*, was published by Monacelli Press, New York, 1994.

5 A short account of Seoul's Live Park can be found online in the e-zine *The Verge* (January 26, 2012) Available at: http://www.theverge.com/2012/1/26/2736462/south-korea-live-park-kinect-rfid-interactive-attractions

6 A celebrated condemnation of the Disney empire can be found in James Howard Kunstler's polemic *The Geography of Nowhere: The Rise and Decline of America's Man-Made Landscape* (Free Press, New York, 1994).

7 A journalistic account of Celebration's dark side can be found in the *Daily Mail* online article by Tom Leonard, titled "The Dark Heart of Disney's Dream Town: Celebration Has Wife-Swapping, Suicide, Vandals . . . and Now Even a Brutal Murder," published on December 9, 2010. Available at: http://www.dailymail.co.uk/news/article-1337026/Celebration-murder-suicide-wife-swapping-Disneys-dark-dream-town.html

8 The National Endowment for the Arts' 2012 Survey of Public Participation in the Arts may be found at: http://arts.gov/publications/highlights-from-2012-sppa.

9 Martin Trondle's eMotion project website contains a wealth of information about the project and is here: http://www.mapping-museum-experience.com/en. Some early results were published with co-authors Steven Greenwood, Volker Kirchberg, and Wolfgang Tschacher in an article titled "An Integrative and Comprehensive Methodology for Studying Aesthetic Experience in the Field: Merging Movement Tracking, Physiology, and Psychological Data," in the journal *Environment and Behavior* (2014, Volume 46, pages 102–135).

10   Dixon published some of the findings I describe with co-authors Kevin Harrigan, Rajwant Sandhu, Karen Collins, and Jonathan Fugelsang in an article titled "Losses Disguised as Wins in Modern Multi-Line Video Slot Machines," in the journal *Addiction*, Volume 105, pages 18–24.

11   Natasha Schull's eye-opening book describes the dramatic changes in the role of slot machines in casinos, *Addicted by Design: Machine Gambling in Las Vegas* (Princeton University Press, Princeton NJ, 2014).

12   Temple Grandin has contributed greatly to our understanding of both autism and animal behavior in a series of technical articles and some great books written for a wide audience. The first of these, describing her remarkable early life and some of her insights about animals, is *Emergence: Labeled Autistic* (Grand Central Publishing, New York, 1996).

13   Bill Friedman's bible of casino design is *Designing Casinos to Dominate the Competition: The Friedman International Standards of Casino Design* (Institute for the Study of Gambling and Commercial Gaming, Las Vegas, 2000).

14   Karen Finlay's group at the University of Guelph have explored the effects of casino design on impulsivity in gambling and the role of gender in an article titled "Casino Décor Effects on Gambling Emotions and Intentions," published in *Environment and Behavior* (2009, Volume 42, pages 542–545).

15   M. Jeffrey Hardwick has explored Victor Gruen's fascinating life in his book *Mall Maker: Victor Gruen, Architect of an American Dream* (University of Pennsylvania Press, Philadelphia, 2003). Malcolm Gladwell wrote about Gruen's influence on American architecture in an article in the *New Yorker* titled "The Terrazzo Jungle," published in the March 15, 2004 issue. Available at: http://www.newyorker.com/magazine/2004/03/15/the-terrazzo-jungle

16   The role of affect in impulse purchasing is described in a 2008 paper by David Silvera, Anne Lavack and Fredric Kropp, titled "Impulse Buying: The Role of Affect, Social Influence, and Subjective Wellbeing," in the *Journal of Consumer Marketing* (2008, Volume 25, pages 23–33).

17   The neurochemistry of impulsivity in rats is described in an article by Marcel van Gaalen, Reinout van Koten, Anton Schoffelmeer and Louk Vanderschuren in an article titled "Critical Involvement of Dopaminergic Neurotransmission in Impulsive Decision Making," in *Biological Psychiatry* (2006, Volume 60, pages 66–73).

18   A wealth of information about Paul Ekman and his current work on the use of facial expression analysis—an approach that he pioneered in a fruitful life work, is described at his website: http://www.paulekman.com/. The Russian Synqera product for using facial expressions for marketing at store checkouts is described at the company website: http://synqera.com/. A short article describing this product can be found on the *Mashable* site in an article written by Adam Popescu on October 2, 2013. Available at: http://mashable.com/2013/10/02/synqera/

Chapter 4

1   Some of the details of my project with the BMW-Guggenheim Laboratory can be found at: http://www.bmwguggenheimlab.org/testing-testing-mumbai

2　Jan Gehl and colleagues Lotte Kaefer and Solvejg Reigstad describe some of their observational studies of the effect of building façades in an article titled "Close Encounters with Buildings," in *Urban Design International* (2006, Volume 11, pages 29–47).

3　This quote from William James comes from page 626 of his *Principles of Psychology, Volume 1* (Henry Holt, New York, 1890).

4　David Berlyne described his "infovore" theory of curiosity and attention in his groundbreaking book *Conflict, Arousal and Curiosity* (McGraw-Hill, New York, 1960).

5　A good accessible but slightly technical account of boredom can be found in John Eastwood's article with colleagues Alexandra Frischen, Mark Fenske, and Daniel Smilek, in a paper titled "The Unengaged Mind: Defining Boredom in Terms of Attention," in the journal *Perspectives on Psychological Science* (2012, Volume 7, pages 482–495).

6　Colleen Merrifield and James Danckert's work on the psychophysiology of boredom is described in a technical article titled "Characterising the Psychophysiological Signature of Boredom," in the journal *Experimental Brain Research* (2014, Volume 232, pages 481–491).

7　Annie Britton and Martin Shipley describe the mortality effects of boredom in an article titled "Bored to Death?" in the *International Journal of Epidemiology* (2010, Volume 39, pages 370–371).

8　The great Canadian psychologist Donald Hebb was decades ahead of his time in describing the influence of experience on brain organization. His original description of his findings with enriched rats can be found in a paper titled "The Effects of Early Experience on Problem-Solving at Maturity" delivered at the 1947 meeting of the American Psychological Association and listed in the journal *American Psychologist* (1947, Volume 2, pages 306–307). His classic book *The Organization of Behavior: A Neuropsychological Theory*, though originally published in 1949 (Wiley and Sons, New York), is still required reading for students of neuroscience—a remarkable thing for such a fast-moving discipline. Mark Rosenzweig described the effects of experience on brain chemistry in anatomy in a book written with Michael Renner titled *Enriched and Impoverished Environments: Effects on Brain and Behavior* (Springer, New York, 1987).

9　Stuart Grassian has written a lengthy report on the psychiatric effects of solitary confinement for the *Washington University Journal of Law and Policy* (2006, Volume 22), available at: http://openscholarship.wustl.edu/law_journal_law_policy/vol22/iss1/24.

10　Aisling Mulligan and colleagues described the influence of the home environment on the development of ADHD in an article titled "Home Environment: Association with Hyperactivity/Impulsivity in Children with ADHD and their Non-ADHD Siblings," in the journal *Child: Care, Health and Development* (2013, Volume 39, pages 202–212).

11　Robert Venturi and colleagues Denise Brown and Steven Izenour described the iconography of urban sprawl in Las Vegas in their controversial book *Learning From Las Vegas* (MIT Press, Cambridge, 1972).

12　Sarah Goldhagen laments the multiple failures of architectural education in an article titled "Our Degraded Public Realm: Multiple Failures of Architecture Education," in the January 10, 2003 issue of the *Chronicle Review*, available on her website (along with many other interesting articles): http://www.sarahwilliamsgoldhagen.com/articles/ multiple_failures_of_architecture_education.pdf

13 Janette Sadik-Khan, New York City's Department of Transportation Commissioner, has made numerous improvements to pedestrian life in the city. The curbside markings are a part of the LOOK campaign, described at: http://www.nyc.gov/html/ dot/html/pr2012/pr12_46.shtml

14 Rem Koolhaas and Bruce Mau describe the Generic City in their book *S, M, L, XL*, (The Monacelli Press, New York, 1997).

15 The quote from Rem Koolhaas comes from an interview with *Der Spiegel* published online and in English on December 16, 2011. Available at: http://www.spiegel.de/ international/zeitgeist/interview-with-star-architect-rem-koolhaas-we-re-building-assembly-line-cities-and-buildings-a-803798.html. The interview was originally published in German in Issue 50 (December 12, 2011) of *Der Spiegel*.

Chapter 5

1 A very interesting and readable account of the Cocoanut Grove fire can be found in John Esposito's book *Fire in the Grove: The Cocoanut Grove Tragedy and Its Aftermath* (Da Capo Press, Boston, 2005).

2 A good review of the relationship between urban social factors and the incidence of psychiatric illnesses can be found in the technical paper by Judith Allardyce and Jane Boydell titled "The Wider Social Environment and Schizophrenia," published in the journal *Schizophrenia Bulletin* (2006 Volume 32, pages 592–598).

3 Several studies have suggested a link between the availability of green space in cities and the incidence of psychosis, depression, and anxiety. Some of the relevant findings are reviewed in the article by Karen McKenzie, Aja Murray, and Tom Booth titled "Do Urban Environments Increase the Risk of Anxiety, Depression and Psychosis? An epidemiological study," published in the *Journal of Affective Disorders* (2013, Volume 150, pages 1019–1024).

4 Florian Lederbogen and a long list of collaborators including Andreas Meyer-Lindenberg published a groundbreaking article on the effects of urban stresses on amygdala activation titled "City living and urban upbringing affect neural social stress processing in humans," in the journal *Nature* (2011, Volume 474, pages 498–501). A readable summary of their findings along with related material can be found in an article by Allison Abbott titled "Stress and the City: Urban Decay," in *Nature* (2012, Volume 490, pages 162–164).

5 Allison Abbot's *Nature* article, cited in the previous note, also provides a short summary of the work by Jim van Os on mental pathology and geotracking.

6 My discussion with Ed Parsons appeared in Land Rover's magazine *Onelife* (2014, Issue #28, pages 40–43) Available at: http://www.landroverofficialmagazine. com/#!parsons-ellard

7 The technical article describing the relationship between neuropeptide S and urban stress was written by Fabian Streit and a large group of collaborators titled "A Functional Variant in the Neuropeptide S Receptor 1 Gene Moderates the Influence of

Urban Upbringing on Stress Processing in the Amygdala," and was published in the journal *Stress* (2014, Volume 17, pages 352–361).

8   Oshin Vartanian describes our preferences for curves and some of its implications for architecture in an article titled "Impact of Contour on Aesthetic Judgments and Approach-Avoidance Decisions in Architecture," in the *Proceedings of the National Academy of Sciences* (2011, Volume 110, Supplement 2, pags 10446–10453) Available at: http://www.pnas.org/content/110/Supplement_2/10446.abstract

9   The experiments describing the effect of geometric shapes on social judgment by Ursula Hess, Orna Gryc, and Shlomo Hareli appear in a paper titled "How Shapes Influence Social Judgments," in the journal *Social Cognition* (2013 Volume 31, pages 72–80).

10   The film *The Pruitt-Igoe Myth*, produced and directed in 2011 by Chad Friedrichs, provides an interesting interpretation of the failure of the development based more on prejudice and economics than on architecture.

11   The dropped letter method was invented by Stanley Milgram (of the infamous Milgram Experiment) and first reported in an article titled "The Lost-Letter Technique: A Tool of Social Research," in the journal *Public Opinion Quarterly* (1965, Volume 29, pages 437–438).

12   The article, titled "Broken Windows: The Police and Neighborhood Safety," that "broke" the news of broken window theory was published in *The Atlantic Monthly* (March, 1982 by James Wilson and George Kelling. In part, their theory was based on earlier work by Philip Zimbardo in an article titled "The Human Choice: Individuation, Reason, and Order Versus Deindividuation, Impulse, and Chaos," in the *Nebraska Symposium on Motivation* (1969, Volume 17, pages 237–307).

13   A report on the Eurobarometer analysis of the fear of crime, produced by the European Commission, titled "Analysis of Public Attitudes to Insecurity, Fear of Crime and Crime Prevention," can be found at: http://ec.europa.eu/public_opinion/archives/ebs/ebs_181_sum_en.pdf

14   A digest of results from a 2010 Gallup poll assessing fear of crime in the United States, titled "Nearly 4 in 10 Americans Still Fear Walking Alone at Night," can be found at: http://www.gallup.com/poll/144272/nearly-americans-fear-walking-alone-night.aspx

15   This Robert Ornstein quote comes from his 1992 book *The Evolution of Consciousness: The Origins of the Way We Think* (Simon and Schuster, New York, page 262).

16   The official Viennese government description of gender mainstreaming may be found here: https://www.wien.gv.at/english/administration/gendermainstreaming/ A good discussion by Clare Foran of the Viennese policies titled "How to Design a City for Women," can be found in the *Atlantic City Lab* blog at: http://www.citylab.com/commute/2013/09/how-design-city-women/6739/

17   The proportion of unmarried adults in U.S. rose to more than 50 percent according to a widely reported survey conducted by the U.S. Bureau of Labor Statistics in 2014. The Martin Prosperity Institute published a regional analysis of the trend in an article written by Richard Florida on September 15, 2014 in the *CityLab* online magazine, titled "Singles Now Make Up More Than Half the U.S. Adult Population. Here's Where They All Live." Available at: http://www.citylab.com/housing/2014/09/singles-now-make-up-more-than-half-the-us-adult-population-heres-where-they-all-live/380137/

18   Statistics from Britain's Office of National Statistics can be found at http://www.ons. gov.uk/ons/rel/census/2011-census-analysis/households-and-household-composition-in-england-and-wales-2001-2011/households-and-household-composition-in-england-and-wales-2001-11.html

19   Statistics on changes in discussion networks in the United States were reported in an article by Miller McPherson, Lynn Smith-Lovin, and Matthew Brashears in an article titled "Social Isolation in America: Changes in Core Discussion Networks Over Two Decades," in the journal *American Sociological Review* (2006, Volume 71, pages 353–375).

20   Findings on loneliness and engagement in Vancouver were reported by the Vancouver Foundation in a 2012 study titled *Connections and Engagement*. Available at: https://www.vancouverfoundation.ca/initiatives/connections-and-engagement

21   A discussion paper by Michael Flood, written for the Australia Institute and titled "Mapping loneliness in Australia" contains a wealth of information about social networks in Australia. Available at: http://www.tai.org.au/documents/dp_fulltext/DP76.pdf A follow-up report, titled "All the Lonely People: Loneliness in Australia, 2001–2009," was published in 2012 by David Baker (also for the Australia Institute) focuses on the issue of loneliness. Available at: http://www.tai.org.au/node/1866

22   John Cacioppo's book, written with William Patrick, is titled *Loneliness: Human Nature and the Need for Social Connection* (Norton: New York, 2009).

23   Keith Hampton's nuanced account of the relationship between electronic social networks and friendship can be found in many of his research articles. A good starting point is his article titled "Core Networks, Social Isolation, and New Media," in *Information, Communication & Society* (2011, Volume 14, pages 130–155).

24   Kevin Bickart and colleagues describe the relationship between amygdala volume and social network size in an article titled "Amygdala Volume and Social Network Size in Humans," in *Nature Neuroscience* (2011, Volume 14, pages 163–164).

25   Keith Hampton and colleagues describe the influence of online social networks on friendship formation in an article titled "How New Media Affords Network Diversity: Direct and Mediated Access to Social Capital Through Participation in Local Social Settings," in the journal *New Media & Society* (2010, Volume 13, pages 1031–1049).

26   John Locke's book is *Eavesdropping: An Intimate History* (Oxford University Press, New York, 2010).

27   Charles Montgomery's book *Happy City: Transforming Our Lives Through Urban Design*, contains a wealth of information about the psychology of urban design. (Farar, Strous & Giroux, New York, 2013).

28   Anthropologist Robin Dunbar first proposed the hard upper limit on social group size in an article on the evolution of neocortex in human beings titled "Neocortex As a Constraint on Group Size in Primates," published in *The Journal of Human Evolution* (1992, Volume 22, pages 469–493). Since this first publication, Dunbar's number has been popularized and applied in a wide range of different contexts.

29   An academic study of the causes of outcries of privacy violation following Facebook's tinkering with newsfeed settings can be found in an article by Christopher Hoadley and colleagues titled "Privacy As Information Access and Illusory Control: The Case

of the Facebook News Feed Privacy Outcry," published in the journal *Electronic Commerce Research and Applications* (2010, Volume 9, pages 50–60).

Chapter 6

1   The audio record of William Anders's comments on Earthrise can be found online as an mp3 file at: http://www-tc.pbs.org/wgbh/amex/moon/media/sf_audio_pop_01b.mp3

2   The quote is taken from MacLeish's essay titled "A reflection: Riders on Earth together, brothers in eternal cold," published in *The New York Times* (December 25, 1968, page 1).

3   The short film *Overview* was produced by Steve Kennedy and the Planetary Collective in 2012. It is available at: http://www.overviewthemovie.com/. Well worth at least one viewing.

4   Dacher Keltner and Jonathan Haidt provide a scientific analysis of the emotion of awe in their 2003 article in the journal *Cognition and Emotion* (2003, Volume 17, pages 297–314).

5   Konrad Lorenz's landmark book *On Aggression* was published in 1963 in German and translated into English in 1966 (Harcourt, Brace and World, San Diego,1966).

6   The study of infant understanding of dominance relationships was published by Susan Carey's group at Harvard in an article titled "Big and Mighty: Preverbal Infants Mentally Represent Social Dominance," in *Science* (2011, Volume 331, pages 477–480).

7   Yannick Joye and Jan Verpooten published an article titled "An Exploration of the Functions of Religious Monumental Architecture from a Darwinian Perspective," describing the evolutionary significance of religious monumental architecture in the journal *Review of General Psychology* (2013,Volume 17, pages 53–68).

8   Laura Kelley and John Endler described perspective illusions in bowerbirds in an article titled "Illusions Promote Mating Success in Great Bowerbirds," in *Science* (2012, Volume 335, pages 335–338).

9   Gordon Gallup described his seminal studies of self-consciousness in primates in an article titled "Self-Recognition in Primates: A Comparative Approach to the Bidirectional Properties of Consciousness," in *American Psychologist* (1977, Volume 32, pages 329–338).

10 Thomas Huxley first laid out his ideas about epiphenomenalism in a paper titled "On the Hypothesis That Animals Are Automata," in the journal *Fortnightly Review* (1874 Volume 16, pages 555–580). A more modern account of his ideas can be found in an article by John Greenwood titled "Whistles, Bells, and Cogs in Machines: Thomas Huxley and Epiphenomenalism," in the *Journal of the History of the Behavioral Sciences* (2010, Volume 46, pages 276–299).

11 Nick Humphrey's fascinating book *Soul Dust: The Magic of Consciousness* provides much food for thought about the biological function of self-consciousness. (Princeton University Press, Princeton NJ, 2011).

12 Ernest Becker's book *The Denial of Death* is scholarly, comprehensive, and deeply disturbing (Free Press, New York, 1973).

13   Terror management theory was first described in a chapter by J. Greenberg,
      T. Pyszczynski, and S. Solomon titled "The Causes and Consequences of a Need for
      Self-Esteem: A Terror Management Theory," in R. F. Baumeister (Ed.), *Public Self and
      Private Self* (Springer-Verlag, New York, pages 189–212). More recent work can be
      found by searching the group's website at: http://www.tmt.missouri.edu/

14   The experiments showing the influence of mortality salience on use of cherished
      symbols can be found in a paper by Greenberg and colleagues titled "Evidence of a
      Terror Management Function of Cultural Icons: The Effects of Mortality Salience on
      the Inappropriate Use of Cherished Cultural Symbols," in the journal *Personality and
      Social Psychology* (1995, Volume 21, 1221–1228).

15   An article describing the influence of mortality salience on George Bush's popularity
      can be found in Mark Landau and colleagues article titled "Deliver Us from Evil: The
      Effects of Mortality Salience and Reminders of 9/11 on Support for President George
      W. Bush," in the journal *Personality and Social Psychology* (2004, Volume 30, pages
      1136–1150).

16   Melanie Rudd's experiments describing the effect of feelings of awe on the perception
      of time can be found in her article with Jennifer Aaker and Kathleen Vohs titled "Awe
      Expands People's Perception of Time, Alters Decision Making, and Enhances Well-
      Being," in the journal *Psychological Science* (2012, Volume 23, pages 1130–1136).

17   Valdesolo and Graham's paper on the relationship between awe and belief in
      supernatural agency titled "Awe, Uncertainty, and Agency Detection," can be found in
      the journal *Psychological Science* (2014, Volume 25, pages 170–178).

18   A good scholarly introduction to body awareness disorders and disruptions of the
      sense of agency can be found in Michela Balconi (ed.), *Neuropsychology of the Sense of
      Agency*, (Springer, New York, 2010).

19   A good account of the rubber hand illusion can be found in the 2007 article by
      Marcello Costantini and Patrick Haggard titled "The Rubber Hand Illusion: Sensitivity
      and Reference Frame for Body Ownership," in the journal *Consciousness and
      Cognition* (2007, Volume 16, pages 229–240).

20   The first description of simulated out of body experiences using virtual reality is given
      by Henrik Ehrsson in a brief note titled "The Experimental Induction of Out-of-
      Body Experiences," in the journal *Science* (2007, Volume 317, pages 1047–1048). An
      elaboration on Ehrsson's procedure is described by Blanke's group in an article titled
      "Video Ergo Sum: Manipulating Bodily Self-Consciousness," in the same volume of
      Science (2007, Volume 317, pages 1096–1099).

21   A comprehensive review of the out-of-body experience, including neuroscientific
      findings, is provided in the chapter by Jane Aspell and Olaf Blanke titled
      "Understanding the Out-Of-Body Experiences from a Neuro-Scientific Perspective,"
      in Craig Murray (Ed.), *Psychological Scientific Perspectives on Out of Body and Near
      Death Experiences* (Nova Science Publishers, Hauppage, NY, pages 73–88, 2009).

22   Fred Previc's fascinating book is *The Dopaminergic Mind in Human Evolution and
      History* (Cambridge University Press, Cambridge, UK, 2009).

23 Previc reviews the role of extrapersonal brain systems in religious activity in an article titled "The Role of the Extrapersonal Brain Systems in Religious Activity," in the journal *Consciousness and Cognition* (2006, Volume 15, pages 500–539).

Chapter 7

1 An article describing the use of virtual reality treatments for post-traumatic stress disorder by Robert McLay and colleagues titled "Effect of Virtual Reality PTSD Treatment on Mood and Neurocognitive Outcomes," can be found in the journal *Cyberpsychology, Behavior, and Social Networking* (2014, Volume 17, pages 439–446).

2 The diary study of mind-wandering and happiness using an iPhone app, conducted by Matthew Killingsworth and Daniel Gilbert, titled "A Wandering Mind Is an Unhappy Mind," can be found in the journal *Science* (2010, Volume 330, page 932).

3 My book *You Are Here: Why We Can Find Our Way to the Moon, but Get Lost in the Mall* was published in 2009 by Doubleday, New York.

4 Nick Yee, Jeremy Bailenson, and colleagues published their analysis of non-verbal behavior in *Second Life* in an article titled "The Unbearable Likeness of Being Digital: The Persistence of Nonverbal Social Norms in Online Social Environments," in the journal *Cyberpsychology and Behavior* (2007, Volume 10, pages 115–121).

5 Edward Hall's influential book *The Hidden Dimension: Man's Use of Space in Public and Private* (The Bodley Head Ltd,. London, 1969).

6 Mel Slater and colleagues' description of bystander responses in virtual reality was published in an article titled "Bystander Responses to a Violent Incident in an Immersive Virtual Environment," in the journal *PLOS One* (2014, Volume 8, article e52766).

7 Blascovitch describes this study and many others in his book with Jeremy Bailenson titled *Infinite Reality: Avatars, Eternal Life, New Worlds, and the Dawn of the Virtual Revolution* (William Morrow, New York, 2011).

8 An article by my student, Kevin Barton, describing his work on wayfinding in virtual environments titled "Seeing Beyond Your Visual Field: The Influence of Spatial Topology and Visual Field on Navigation Performance," can be found in the journal *Environment and Behavior* (2012, Volume 46, Pages 507–529).

9 An entertaining and informative description of the Palmer Luckey story by Taylor Clark titled "How Palmer Luckey Created Oculus Rift" can be found in the issue of *Smithsonian Magazine* (November, 2014). Available at: http://www.smithsonianmag.com/innovation/how-palmer-luckey-created-oculus-rift-180953049/?no-ist

10 Much illuminating information about the video game industry can be found at the website of the Entertainment Software Rating Board. Available at: http://www.esrb.org/about/video-game-industry-statistics.jsp

11 An early and entertaining account of teledildonics titled "Teledildonics: Reach Out and Touch Someone" was written by Howard Rheingold in the journal *Mondo 2000* (Summer, 1990). The article, complete with an opening cheeky limerick, is available in the book Arthur Berger (Ed.), *The Postmodern Presence: Readings on Postmodernism in American Culture and Society* (Rowman Altamira, New York, 1998).

12 Project Syria and the virtual Aleppo experience is described at the Immersive Journalism website at: http://www.immersivejournalism.com/

13 The article by Rosenberg, Baughman and Bailenson describing the virtual superhero effect, titled "Virtual Superheroes: Using Superpowers in Virtual Reality to Encourage Prosocial Behavior," was published in the journal *PLOS One* (2013, Volume 8, pages 1–9).

14 George Stratton's description of the inverting goggles experiments can be found in an ancient paper titled "Some Preliminary Experiments on Vision Without Inversion of the Retinal Image," in the journal *Psychological Review*, (1896, Volume 3, pages 611–617). A somewhat more recent review written by Charles Harris titled "Perceptual Adaptation to Inverted, Reversed, and Displaced Vision," describes the research of Stratton and many others on the perceptual consequences of inverted and reversed vision. Harris's paper can be found in *Psychological Review* (1965, Volume 72, Pages 419–444).

15 William Warren's work with Jonathan Ericson on wormholes in virtual reality, titled "Rips and Folds in Virtual Space: Ordinal Violations in Human Spatial Knowledge," can be found in abstract form in the archive of the online *Journal of Vision* (2009, volume 9, article 1143). Available at: http://www.journalofvision.org/content/9/8/1143. meeting_abstract

16 The essay by Walter Benjamin was first published in 1936 in French in the journal *Zeitschrift für Sozialforschung* (1936, Volume 5, pages 40–68). An English translation can be found at: http://www.marxists.org/reference/subject/philosophy/works/ge/ benjamin.htm

Chapter 8

1 The original paper on calm technology, titled "The Coming Age of Calm Technology," was published by Marc Weiser and John Seely Brown in the journal *PowerGrid* (1996, Volume 1.01). A revised version is available on Seely Brown's website at: http://www. johnseelybrown.com/calmtech.pdf

2 Dan Hill's essay, titled "The Street as Platform," can be found on his blog *City of Sound* at: http://www.cityofsound.com/blog/2008/02/the-street-as-p.html

3 If you enjoy infographics, take a look at Google's consumer barometer at: https://www. consumerbarometer.com/en/. Many more interesting statistics are available in a report from Nielsen, available at: http://www.nielsen.com/content/dam/corporate/us/en/ reports-downloads/2013%20Reports/Mobile-Consumer-Report-2013.pdf

4 An interesting comparison of the power of an iPhone with the computing power of rocketships, titled "A Modern Smartphone or a Vintage Supercomputer: Which Is More Powerful?" can be found at: http://www.phonearena.com/news/A-modern-smartphone-or-a-vintage-supercomputer-which-is-more-powerful_id57149.

5 A technical account of the roles of different brain areas in route-following and map-based wayfinding can be found in the article by Tom Hartley, Eleanor Maguire, Hugo Spiers, and Neil Burgess titled "The Well-Worn Route and the Path Less Traveled: Distinct Neural Bases of Route Following and Wayfinding in Humans," in the journal *Neuron* (2003, Volume 37, pages 877–888).

6 John O'Keefe and Lyn Nadel's groundbreaking book *The Hippocampus as a Cognitive Map* was first published in 1978 by (Clarendon Press, Oxford, UK). The book is out of print but available online at: http://www.cognitivemap.net/HCMpdf/HCMComplete. pdf. O'Keefe shared the 2014 Nobel Prize in Physiology or Medicine with Edvard and

May-Britt Moser, who have worked out many more of the details underlying spatial mapping in brains. They tell the story in their own words at: http://www.ntnu.edu/kavli/discovering-grid-cells.

7   Eleanor Maguire's team describe some of the fascinating findings related to hippocampal structures and spatial navigation in London taxi drivers in an article titled "Navigation-Related Structural Change in the Hippocampi of Taxi Drivers," in the journal *Proceedings of the National Academy of Sciences (USA)* (2000, Volume 97, pages 4398–4403).

8   Véronique Bohbot's group describe changes in navigation strategies across the lifespan and their implications for healthy aging in an article titled "Virtual navigation strategies from childhood to senescence: Evidence for changes across the life span," in *Frontiers in Aging Neuroscience* (2012, Volume 4, Article 28). An account of their earlier findings suggesting that GPS use may reduce hippocampal function can be found at the Phys.org website at: http://phys.org/news/2010-11-reliance-gps-hippocampus-function-age.html

9   Albert Borgmann's important and brilliant book *Technology and the Character of Contemporary Life* rewards close reading, (University of Chicago Press, Chicago, 1985).

10   Daniele Quercia's group at Yahoo describe their work in an article titled "The Shortest Path to Happiness: Recommending Beautiful, Quiet, and Happy Routes in the City," in the *Proceedings of the 25th ACM conference on Hypertext and Social Media*, (ACM Press, New York, pages 116–125, 2014). The article is also available at: http://dl.acm.org/citation.cfm?id=2631799

11   Sadly, the MATR app is no longer available, but some details of the project are available at: http://www.spurse.org/what-weve-done/matr/

12   Bruce Sterling's provocative pamphlet, titled *The Epic Struggle of the Internet of Things* (Strelka Press, Moscow, 2014).

13   The story of the use of Fitbit evidence in a personal injury suit is recounted by Kate Crawford in a November 19, 2014 article titled "When Fitbit is the Expert Witness," in *The Atlantic*. Available at: http://www.theatlantic.com/technology/archive/2014/11/when-fitbit-is-the-expert-witness/382936/

14   Adam Greenfield's book *Against the Smart City* (Do Projects, New York, 2013) was published as an e-pamphlet. Available at: Amazon.

Conclusions

1   IDIA Lab's Virtual Stonehenge project may be found at: http://idialab.org/virtual-stonehenge/

For Further Reading

I provide this somewhat idiosyncratic list of some of my favorite books on topics related to those I've covered in this book. For one reason or another, these books did not find their way into the footnotes but all reward close reading.

De Botton, Alain. *The Architecture of Happiness*. New York: Pantheon, 2006.

Eberhard, John P. *Brain Landscape: The Coexistence of Neuroscience and Architecture*. Oxford: Oxford University Press, 2008.

Hildebrand, Grant. *Origins of Architectural Pleasure*. Oakland, CA: University of California Press, 1999.

Marcus, Clare Cooper. *House as a Mirror of Self: Exploring the Deeper Meaning of Home*. Newburyport, MA: Red Wheel, 2002.

Mallgrave, Harry F. *The Architect's Brain: Neuroscience, Creativity and Architecture*. Hoboken, NJ: Wiley & Sons, 2009.

Pallasmaa, Juhani. *The Eyes of the Skin: Architecture and the Senses*. Hoboken, NJ: Wiley & Sons, 1996.

Pollan, Michael. *A Place of My Own: The Education of an Amateur Builder*. New York: Random House, 1997.

Rasmussen, Steen Eiler. *Experiencing Architecture*. Cambridge, MA: The MIT Press, 1995.

Sternberg, Esther M. *Healing Spaces: The Science of Place and Well-Being*. Cambridge, MA: Harvard University Press, 2010.

Zeisel, John. *Inquiry By Design: Environment/Behavior/Neuroscience in Architecture, Interiors, Landscape, and Planning*. New Yor: W. W. Norton, 2006.

Index

Acknowledgments

M Y LESSONS IN PSYCHOGEOGRAPHY began early in life with a great set of boyhood friends who tramped with me through forests, fields, suburban streets, and late-night cityscapes where we learned together most of what is worth knowing about place, feelings, and perhaps most importantly, how to get into and back out of trouble. For those lessons I thank Mark Wright, Steve Mason, David Westlake, Vince Burry, George Danes, and Paul Ruttan.

My more formal mentors, who taught me to love neuroscience and helped me to imagine a future in the field, include John Yeomans, Mel Goodale, Case Vanderwolf, and Barrie Frost. Thomas Seebohm introduced me to the science, art, and mystery of architecture and also to some of his amazing colleagues, including Philip Beesley, whose insight, imagination, and passion have had more of an impact on me than he could possibly know.

I thank the group from the BMW-Guggenheim Laboratory for providing me with unprecedented opportunities to conduct some global psychogeography—Charles Montgomery who got me into the whole business, David Van der Leer and Maria Nicanor, who supported and encouraged my experiments, and countless staff, interns, and volunteers, without whom I could not have done experiments in fascinating and far-reaching places. In particular, Stephanie Kwai, Christine MacLaren, Andres Carter, Alexander Bolinder-Gibsand, Constantin Boese, and Mahesh Makwana provided tireless hours of support. Most of all, I thank the hardy volunteers who followed me on psychogeographic walks through oppressive heat, torrential rain, and teeming crowds of curious onlookers.

A special thank you to Ruth and John Corner, and their son Ian, who graciously permitted me to visit one of my earliest homes in England and so galvanized many of my thoughts about home.

On the home front, I thank colleagues, staff, and students at the University of Waterloo who have contributed to some of the work that I describe here. Without the enthusiasm and long hours of toil of graduate students like Deltcho Valtchanov, Kevin Barton, Vedran Dzebic, Kaylena Ehgoetz Martens, and Hanna Negami, nothing much of any importance can get done in a busy research laboratory.

I thank my children, whose own early forays into the art of psychogeography have amused and inspired me, and my father, Ronald, who, among so many other things, first helped me to imagine what it means to build something.

I'm very grateful to my agent, Denise Bukowski, who worked hard to help me find a good home for this book, and my editor at Bellevue Literary Press, Erika Goldman, whose enthusiasm, acuity, wisdom, and great sense of humor helped me to get to the finish line.

Finally, I thank Kristine, whose love and support sustains me every day of my life and inspires me to do my best and to keep reaching further.